The Politics of Knowledge

Edited by
Fernando Domínguez Rubio and
Patrick Baert

LONDON AND NEW YORK

First published 2012
by Routledge
2 Park Square, Milton Park, Abingdon, Oxon OX14 4RN

Simultaneously published in the USA and Canada
by Routledge
711 Third Avenue, New York, NY 10017

Routledge is an imprint of the Taylor & Francis Group, an informa business

British Library Cataloguing in Publication Data
A catalogue record for this book is available from the British Library

Library of Congress Cataloging in Publication Data
A catalog record has been requested for this book

First issued in paperback 2013

ISBN: 978-0-415-70475-5 (pbk)
ISBN: 978-0-415-49710-7 (hbk)
ISBN: 978-0-203-87774-6 (ebk)

Typeset in Baskerville
by Wearset Ltd, Boldon, Tyne and Wear

Contents

Contributors

Patrick Baert is Fellow of Selwyn College and Reader in Social Theory at the University of Cambridge.

Ulrich Beck is Emeritus Professor of Sociology at the University of Munich and British Journal of Sociology Visiting Centennial Professor at the London School of Economics.

Fernando Domínguez Rubio is a Postdoctoral Marie Curie Fellow at New York University and the Centre for Research on Socio-Cultural Change at The Open University.

Fernando J. García Selgas is Professor at the Universidad Complutense de Madrid.

Sheila Jasanoff is Pforzheimer Professor of Science and Technology Studies at Harvard University.

John Law is Professor of Sociology at the Open University and Director of the Centre for Research on Socio-Cultural Change.

James Leach is Professor in Social Anthropology at the University of Aberdeen.

Javier Lezaun is James Martin Lecturer in Science and Technology Governance and a Fellow at Kellogg College.

Saskia Sassen is Robert S. Lynd Professor of Sociology, Columbia University, Committee on Global Thought, Columbia University and Centennial Visiting Professor, London School of Economics.

Alan Shipman is Lecturer in Economics at the Open University.

Bryan S. Turner is Presidential Professor at the Graduate Center, City University of New York.

Peter Wehling is Research Fellow in Sociology at the University of Augsburg.

Politics of Knowledge

An introduction

Fernando Domínguez Rubio and Patrick Baert

Over recent decades it has become commonplace to refer to contemporary advanced societies as 'knowledge societies'. This indicates the degree to which different forms of knowledge production – like science and technology – and distribution – like information and communication technologies – are now fundamental processes in the fabric of advanced societies. Be it in the form of communication devices, transport systems, domestic technologies, energy infrastructures, medical or economic expertise, our contemporary way of life has become dependent, more than at any previous time, on a wide variety of technical and scientific knowledge. Simultaneously, the unabated innovation and diffusion of digital communication and information has increased to an unprecedented extent the potential for the production and distribution of knowledge. The rapid expansion of these technologies – together with their instantaneity and transnational nature – has resulted in what arguably constitutes the first truly global networks and flows of knowledge and information. Moreover, as economic competitiveness and productivity have become more dependent on research and innovation, knowledge production and distribution have become increasingly central processes in the generation of value in contemporary capitalist economies. In this rapidly evolving knowledge-intensive context, the social sciences have produced a thriving body of scholarly work dealing with different aspects of contemporary knowledge production and distribution. One focus of this literature has been on the new modes of techno-scientific knowledge production (Gibbons *et al.* 1994) as well as the different 'epistemic cultures' (Knorr-Cetina 1999), material cultures (Galison 1997), practices (Latour 1987; Pickering 1995) and forms of legitimacy (Daston and Galison 2007), which lie behind them. Another focus of this literature has been on the relations between knowledge and the economy; in particular, the transformative effect of knowledge in the working logic of Western economies and financial markets, and the rise of the so-called knowledge-intensive economies (Thurow 2000; Adler 2001; Chichilnisky and Gorbachev 2004; Powell and Snellman 2004). Although the relations between knowledge, science and economy have benefited from this growing current of scholarly interest, much less attention has been paid to the new and intricate relationships that are developing in this context between knowledge and politics. How do these new dynamics of knowledge production and distribution affect established political categories and boundaries? Which political vocabularies and institutions are required to govern these novel forms of knowledge production and distribution? And, crucially: what are the political opportunities and risks emerging from these processes? These questions constitute the central focus of this book.

This volume brings together the work of a number of scholars who, in a variety of ways, try to reconceptualise the relationship between politics and knowledge. The title of this book, *Politics of Knowledge*, captures what they have in common: the recognition that knowledge is constitutive of the world and therefore political. In this respect, the authors argue against a widespread and influential notion of the relationship between knowledge and politics which, for the sake of brevity, will be referred to henceforth as the 'liberal view'. The central tenet of this liberal view is that knowledge and politics are, and must be kept as, separate activities (Merton 1979; Weber 2004; see also Jasanoff, Chapter 1, this volume). Thus, while politics should concern itself with the sphere of values – with how the world *ought* to be – knowledge should be exclusively concerned with the sphere of facts – with how the world *is*. Knowledge, it follows, should be regarded as a mirror that passively registers, without interfering, the essential features and causal relations already existing in the world. To put it differently, knowledge is, and ought to be, value-free, objective and, therefore, apolitical. Any knowledge interfered with or tinged by politics ceases to be knowledge, and becomes mere ideology – a value-ridden representation of the world. Politics, on the other hand, should place an equal emphasis on curtailing the spectre of technocracy that constantly threatens to reduce free political debate to the tyrannical rule of experts. The only legitimate interaction between knowledge and politics is, according to this liberal view, that by which knowledge provides the means for achieving effectively and efficiently the goals democratically set in the political sphere. Although neither knowledge nor politics might, in practice, fulfil the lofty ambitions postulated by this liberal understanding, the ideal of this separation has been long held up as the yardstick or normative ideal by which both activities should be practised and judged. That is, even if the proponents of this liberal view accept that it is impossible to disentangle knowledge production from the political context in which it takes place, they nonetheless postulate that knowledge production should *strive* to be as free as conceivably possible from any political interference that could taint the ideals of objectivity and universality. Likewise, even if in practice political debate is always exposed to different forms of expert knowledge and vested interests, political debate should endeavour to attain an 'ideal speech situation' in which moral and political concerns can be freely advanced and defended by rational means alone (Habermas 1992).

It could be argued that this liberal ideal of an orderly division of labour between knowledge and politics provided a reasonable and productive normative framework for much of the twentieth century. Our argument in this book is that the transformations in the regimes of knowledge production and distribution taking place over recent decades have rendered this view increasingly unproductive for understanding the new relations between knowledge and politics, and indeed for providing a normative yardstick to judge both activities. In contrast to the subservient role envisaged for knowledge in this liberal view, knowledge has today become a source of questions that need to be politically answered, rather than just a means to answer political questions. Examples abound. Recent developments in neurosciences, reproductive technologies, human enhancement technologies and genetic engineering have rendered uncertain and open to debate some of the previously incontrovertible biological foundations on which modern juridico-political categories have been built (Haraway 1997; Rose 2006; Franklin 2007). The traditional identification of political subjects with the biological body, for instance, is challenged by the emergence of technologically enhanced bodies which inhabit transitional zones in which their political status, as well as the nature and extent of their civil and political rights, remains uncertain. Similarly, the emergence of genetically modified organisms calls into question the traditional separation between nature, technology and culture, as well as a host of categories, like property or indeed the very definition of life itself, which were erected upon them. The development of new communication and information technologies, on the other hand, defies conventional hierarchies of expertise as well as the institutions through which knowledge has been customarily produced, circulated and legitimated (Latham and Sassen 2005). By enabling the creation of communities and bodies of knowledge that exist beyond the boundaries and control of hierarchical institutions, these technologies open the door to forms of political association and action which unsettle the traditional identification of the political with the formal structures of the state apparatus. In addition to these developments, the large-scale incorporation of knowledge into the market has resulted in knowledge-intensive economies which have subjugated knowledge production to market demands and agendas. As a result of this, a novel pragmatics of knowledge has emerged in which truth value has been gradually replaced by productivity and exchange value. This growing commodification of knowledge has radically transformed the old institutional apparatus of knowledge production, with universities and different research enterprises increasingly concerned with catering for markets by producing 'useable' and 'valuable' knowledge (Strathern 2000).

These different, albeit parallel developments point in a direction that is in open contrast with that optimistic Enlightenment ideal which saw knowledge as a civilising force aimed at the betterment of human societies. Far from the idea of knowledge as a cumulative process that gradually advanced by dispelling previous areas of ignorance, contemporary knowledge has emerged, instead, as a Janus-faced phenomenon in which every piece of new knowledge is invariably accompanied by uncertainties and risks which have to be politically governed and managed. This Janus-faced nature is evident, for example, in the burgeoning areas of genetic engineering and novel reproductive technologies. Although these techno-scientific knowledges have opened up a promising prospect of curing genetic and chronic diseases, they have also produced an equally impressive number of daunting ethical and political debates that remain unresolved. It is still unclear what are the limits to human intervention and manipulation of life or how such limits could be implemented. Similarly, and despite intensive investigation, there is as yet no definitive answer about the long-term health and environmental effects of spreading genetically modified foods. Developments like man-made climate change, the recent crash of knowledge-intensive financial markets, or recurrent food and health pandemics, painfully reveal the extent to which knowledge and its applications can produce unanticipated, even disastrous, effects. If anything has been learned over the recent decades of intensive knowledge production it is that this penumbra of uncertainties and risks is not due to temporal deficiencies in knowledge – which will be eventually washed away by further discovery – but is, in fact, a constitutive part of contemporary knowledge societies. It is this that makes it possible to speak of 'the paradox of knowledge' in contemporary knowledge societies. Far from the opposition between knowledge and ignorance envisaged by Enlightenment thinkers, knowledge-intensive societies face the paradoxical, and seemingly endless multiplication of knowledge-manufactured uncertainties. It is this co-production of knowledge and non-knowledge, we argue, that constitutes one of the key defining features of contemporary societies (see Beck and Wehling in this volume).

In this context, it is not surprising to observe that discussions about knowledge have gradually shifted away from old epistemological debates about the 'representational capacity of knowledge' which dominated much of the twentieth century. Indeed, the old tugs-of-war between the 'positivist' defenders of universal and objective knowledge and different varieties of 'relativist' have been slowly replaced by a growing concern about the ontological dimensions of knowledge; that is, about its *generative capacity* to produce new entities and relations in the world. In other words, in the age of bio-technology, new communication technologies and information economies, the key challenge is no longer to determine the 'truth conditions' of these knowledges, or whether they actually provide an objective and pure representation of reality. The most urgent task, instead, is to discuss the possibilities, risks and effects of the new realities generated by these novel forms of knowledge production and distribution, and also the political and legal settings required to accommodate and govern these realities. In a nutshell, the key question at the dawn of the twenty-first century is the politics of knowledge. This volume is a first attempt to address the contemporary politics of knowledge.

Although this book approaches the politics of knowledge in different ways, it is possible to distinguish four distinct sub-themes. The first of these sub-themes explores the institutional frameworks and vocabularies required to govern the paradoxical co-production of knowledge and non-knowledge which characterises contemporary societies. Here the key questions are: how can we think anew about the relationship between knowledge and politics, and about that between non-knowledge and politics? Two chapters in this volume – Chapter 2 by Ulrich Beck and Peter Wehling, and Chapter 1 by Sheila Jasanoff – tackle these questions. Jasanoff's chapter urges us to rethink the relationship of knowledge and politics beyond the strict separation between these two realms postulated by what we have called the 'liberal view'. In order to do so, Jasanoff redefines scientific knowledge and technological applications as 'agents of political production' with repercussions for how we imagine and implement democracy. Science, she contends, is political 'all the way down'. Jasanoff's project to 'repoliticise' science requires a novel understanding of the ways in which the public can participate in the governance of scientific knowledge. Distancing herself from the 'public understanding of science' (PUS) model that has been the hegemonic framework to understand public engagement in the governance of science, Jasanoff offers us an innovative model based on 'knowledge rights'. According to Jasanoff different forms of knowledge rights, like the right to know, to give informed consent or to demand reasons, have in fact been deployed over recent decades in the form of new legislation and regulation, such as freedom of information laws, consumer protection laws or rules of administrative procedure. In contrast to the PUS model which tended to disempower citizens by portraying them as a mass of 'ignorant children', Jasanoff argues that the 'knowledge rights' encoded in these laws constitute a tool to empower willing citizens to participate in the governance of science and, more widely, a necessary step towards building democracies of reason.

If Jasanoff's chapter offers as a way of rethinking the governance of knowledge, Beck and Wehling's contribution draws our attention to the other side of the paradox; that is, the governance of non-knowledge. Their specific focus here is on what they call the 'politicisation of non-knowing'. As they contend, 'non-knowledge' is rapidly becoming a site of political struggle in which two interpretative modes clash. Whereas some people insist, following the Enlightenment tradition, that non-knowing is only a temporary problem and that causal networks will eventually be established, others argue that we should acknowledge the constitutive nature of non-knowledge and therefore build this non-knowing into the way we approach the world. Drawing evidence from the recent global financial crisis, debates about climate change, genetically modified food and predictive genetic testing, the authors show the increasing pervasiveness of non-knowledge in contemporary knowledge practice and the need to develop new forms of decision-making that go beyond the modernist assumption that non-knowing is always merely temporal. Non-knowledge, Beck and Wehling argue, is here to stay; hence the urgent need to incorporate it as an integral part of our political vocabularies.

Continuing with the exploration of the uncertainties and possibilities emerging from the dynamics of contemporary knowledge, the second subtheme of the book focuses on how these dynamics are giving rise to forms of subjectivity and objectivity which call into question some long-standing political and legal categories. Fernando Domínguez Rubio and Javier Lezaun's contribution (Chapter 3) is an example of how new technologies create forms of personhood that unsettle biological demarcations that have hitherto been used to define political categories like 'citizenship' or the 'political subject'. Starting from two case studies of Locked-in-Syndrome (LIS) patients, Domínguez Rubio and Lezaun suggest that new forms of knowledge and technological enhancement applied to these patients are giving rise to 'distributed forms of personhood' in which various capacities associated with personhood (such as agency or speech) are delegated on to and brought about by a combination of biological, technical and social elements. The authors analyse the difficulties of legal knowledge in recognising these emerging distributed forms of personhood as full-fledged political subjects and the collective effort required to produce and sustain the conditions of intelligibility that LIS patients need to become known as full-blown citizens. Through this analysis the authors seek to advance a novel socio-material perspective on citizenship that regards it as a fragile 'position' embedded in socio-technical systems of knowledge and care, rather than as an abstract 'condition' grounded in an isolated, self-governing body.

In a similar vein, Leach (Chapter 4) also focuses on different legal categories, like cultural and intellectual property, to trace the emergence of alternatives to dominant understandings of knowledge. The rise of knowledge economies and their array of commodification and auditing practices has resulted, according to Leach, in the reification of knowledge as a discrete and tradable 'object'. It is this focus on discrete objects, rather than on social relations, that has led notions of cultural and intellectual property to neglect modes of knowledge and value production that escape the logic of quantification and commodification. Drawing data from ethnographic studies with artists in Indonesia and Melanesia and a collaborative research enterprise between scientists and artists in the UK, Leach demonstrates the possibility of a mode of knowledge production in which value does not derive from the capacity of knowledge to be transacted as an object in the market, but from its capacity to generate specific 'effects' on social relations and persons. This focus on effects, rather than on objects, Leach argues, opens a door to escape the dominant "false scale of accounting in which comparative judgements about value are made to the detriment of recognising wider diverse, social benefits".

The third sub-theme in this volume explores how information technologies, partly because of the extent to which they create new forms of knowledge distribution, can undermine established political structures and create new possibilities for political action. Saskia Sassen (Chapter 5) discusses how the new interactive information and communication technologies can give rise to new types of economic structures and 'informal politics'. Sassen analyses different contexts in which the new technologies can be and are being used; these contexts range from contemporary global financial markets to the Zapatistas' intensive political struggles. These examples, both in the economic and the political realm, show new, complex ways in which the local and the global interact. In her discussion of political phenomena, Sassen shows that the new technologies do not necessarily guarantee democratic processes, just as it would be wrong to say that they *ipso facto* impede democratic potential. Rather, Sassen's discussion indicates that the new computer-based networks and the increasing digitisation of knowledge, if directed and implemented properly, have the capacity to establish new political forms of mobilisation, some of which indeed constitute an expansion of non-hierarchical political processes.

Bryan Turner (Chapter 6) complements Sassen's contribution by highlighting the role of the new information technologies in eroding traditional hierarchies. Specifically, Turner explores the effect of increasing literacy levels, democratisation of knowledge and widespread access to new technologies on religious knowledge and authority. Drawing on a wide variety of empirical sources, Turner argues convincingly that, in different parts of the world, the 'age of revelation' has been substituted for the 'age of information'. Whereas the former takes essential truth to be arcane and essentially hidden, the latter sees truth as basically accessible to all. This, according to Turner, is what secularisation is really about: the blurring of the distinction between the elite and the masses eventually undermines 'hierarchically organised wisdom' and the control over (or even the very notion of) the 'ineffable'. In contemporary knowledge-based societies, the hitherto ineffable divine messages transmitted by theologians or priests are disseminated by modern "intermediaries (talk-show hosts, opinion leaders, journalists, TV personalities and the like) who make the ineffable effable". As Turner argues, the transformation enacted by these new systems of knowledge production and distribution explains to a great extent the survival, if not buoyancy, of religion in contemporary Western as a form of "spirituality which is a post-institutional, hybrid and individualistic religiosity".

We did not want to conclude our exploration of the contemporary politics of knowledge without investigating the politics of the types of knowledge produced by those who investigate and participate in the social world. This is the topic of the fourth sub-theme, which explores the positions and tools that social scientists may employ to deal with this complex and evolving relation between knowledge and politics. Fernando

García Selgas (Chapter 7) tackles this issue head-on by proposing a paradigm of social fluidity. For almost three decades now, Selgas argues, social theorists have tried to study social life in terms of fluidity, whether this is expressed as 'liquidity', 'fragmentation', 'mobility' or 'relationality'. This new paradigm of fluidity brings together some otherwise very different orientations, ranging from Baudrillard's reflections on media-driven society and Zygmunt Bauman's observations about liquid modernity to Bruno Latour's attempts to bypass the rigid opposition between nature and culture. Selgas argues that these new theoretical developments towards recognising fluidity undermine previously stable conceptual entities like the state or the political subject and open the door to a new kind of 'fluid politics' which enacts alternative forms of political discourses and practices. Rather than politics based on discrete and fixed entities (like the state or the subject), this model of fluid politics points in the direction of a vocabulary rooted in changing, heterogeneous entities "more interconnected and implicated with the environment and other voices".

John Law's work (Chapter 8) falls squarely within this paradigm of fluidity. In his contribution, Law focuses on the performative effects that our methods have on the realities we study. He introduces the notion of 'collateral realities' to refer to those realities that are accomplished, mainly unintentionally, as we frame and study reality. By developing this approach, Law opposes what he sees as 'Euro-American common-sense realism' which assumes that there is a coherent, single reality that exists independently of people's action. Countering this view, Law claims that realities are 'done' and 'accomplished' through a variety of practices that involve different sets of 'material–semiotic relations'. Law spells out this view through a particular example: a stakeholders' meeting of a programme called Welfare Quality on farm animal welfare. By focusing on the specific methods and technologies employed in this meeting to produce knowledge claims about animal welfare, Law shows how methods and knowledge perform, along the way, specific understandings of the citizen, the consumer and the state as 'collateral realities'. Yet, as Law warns, this view of reality as being done by different sets of practices does not amount to saying that 'anything goes'. Rather, he claims: "it is to shift our understanding of the *sources* of the relative immutability and obduracy of the world: to move these from 'reality itself' into the choreographies of practice." It is in this sense, Law argues, that it is possible to define knowledge practices as a form of ontological politics; that is, a way of enacting specific realities as well as a host of unintended, but equally crucial collateral realities.

Still within this fourth sub-theme, Patrick Baert and Alan Shipman (Chapter 9) discuss the phenomenon of public intellectuals and argue against the view that their status and number are in decline. Against this 'declinist' thesis, Baert and Shipman assert that new types of public engagement have emerged which also result in novel forms of political engagement. Commentators who advocate the declinist thesis tend to take too restrictive a notion of what it is to be an intellectual, and fail to recognise the new forms of, and new participants in the intellectual sphere. The prototypical cases were authoritative intellectuals: generalists with a considerable amount of cultural capital and a certain aura, often taking a moral stance. Authoritative intellectuals have gradually been replaced by professional intellectuals, and then by what are termed embedded intellectuals. In contrast with authoritative intellectuals, professional intellectuals are steeped in a particular discipline and derive their authority from that expertise. However, like authoritative intellectuals, professional intellectuals speak from above, whereas embedded intellectuals have a more democratic relationship with their audience, often developing a dialogue with a wider public and relying on it to boost their credibility and survival.

Although the four sub-themes do not represent an exhaustive list, they are likely to play a central role in research into the politics of knowledge for the foreseeable future. As such, the study of the politics of knowledge, anchored as it is in those four sub-themes, has major repercussions for various disciplines other than sociology and political science while also drawing on those other disciplines to reshape sociological and political concepts. For instance, while Turner's chapter shows the significance of new technologies for the study of religion, Leach's contribution exemplifies the importance of these issues for scientists and artists, and across different societies; and Domínguez Rubio and Lezaun's arguments about the subject and personhood explore emerging issues in legal theory. These examples demonstrate how the study of the politics of knowledge invites a truly cross-disciplinary approach that not only opens up dialogue between neighbouring disciplines in the social sciences but possibly calls into question the neat divisions between them. Furthermore, several contributions, including Sassen, García Selgas, Law, and Domínguez Rubio and Lezaun, emphasise the contemporary importance of hybrid entities that shatter the separation between the social and the technological. This volume is, therefore, an invitation to rethink disciplinary as well as ontological boundaries.

The issues raised in this volume are topical and will lastingly shape the nature of politics and society. But it would be misleading to take academics, like those contributing to this book, as having a monopoly over the type of knowledge that the book identifies and assesses. There are many others who reflect on the politics of knowledge, especially in an age where, as Baert and Shipman point out, sections of the highly educated public have plenty of resources and time at hand to develop sophisticated analyses and criticisms. Journalists, politicians, scientists, policy makers, artists and activists regularly analyse a substantial number of the topics discussed here. In an era of reflexive politics, considerations similar to those we find in this book feed back into the political sphere and will eventually change the future political direction. The chapters by Jasanoff and Beck and Wehling give some indication of the extent to which reflections on the politics of knowledge are already woven into the fabric of society today, but more of those feedback loops are to come, making the future of the politics of knowledge both exciting and unpredictable.

Bibliography

Adler, Paul S. 2001. Market, Hierarchy, and Trust: The Knowledge Economy and the Future of Capitalism. *Organization Science* 12, no. 2: 215–234.

Chichilnisky, Graciela and Olga Gorbachev. 2004. Volatility in the Knowledge Economy. *Economic Theory* 24, no. 3: 531–547.

Daston, Lorraine and Peter Galison. 2007. *Objectivity*, 1st edn. New York: Zone Books.

Franklin, Sarah. 2007. *Dolly Mixtures: The Remaking of Genealogy.* Princeton, NJ: Duke University Press Books.

Galison, Peter Louis. 1997. *Image and Logic: A Material Culture of Microphysics.* Chicago, IL: University of Chicago Press.

Gibbons, M., Camille, L., Schwartzman, S., Nowotny, H., Trow, M. and Scott, P. 2004. *The New Production of Knowledge: The Dynamics of Science and Research in Contemporary Societies.* Sage: London.

Habermas, Jurgen. 1992. *Moral Consciousness and Communicative Action.* Boston, MA: MIT Press, October 8.

Haraway, Donna J. 1997. *Modest_Witness@Second_Millennium.FemaleMan_Meets_OncoMouse: Feminism and Technoscience.* London: Routledge, 15 January.

Knorr-Cetina, K. 1999. *Epistemic Cultures: How Sciences Make Knowledge.* Cambridge, MA: Harvard University Press.

Latham, Robert, and Saskia Sassen. 2005. *Digital Formations: IT and New Architectures in the Global Realm.* Princeton, NJ: Princeton University Press.

Latour, Bruno. 1987. *Science in Action: How to Follow Scientists and Engineers through Society.* Boston, MA: Harvard University Press.

Merton, Robert K. 1979. *The Sociology of Science: Theoretical and Empirical Investigations.* Chicago, IL: University of Chicago Press.

Pickering, Andrew. 1995. *The Mangle of Practice: Time, Agency, and Science.* Chicago, IL: University of Chicago Press, 15 August.

Powell, Walter W. and Kaisa Snellman. 2004. The Knowledge Economy. *Annual Review of Sociology* 30, no. 1: 199–220. doi:10.1146/annurev.soc.29.010202.100037.

Rose, Nikolas. 2006. *The Politics of Life Itself: Biomedicine, Power, and Subjectivity in the Twenty-First Century*, annotated edn. Princeton, NJ: Princeton University Press.

Strathern, Marilyn. 2000. *Audit Cultures: Anthropological Studies in Accountability, Ethics and the Academy.* London: Routledge, 8 September.

Thurow, Lester C. 2000. Globalization: The Product of a Knowledge-Based Economy. *Annals of the American Academy of Political and Social Science* 570: 19–31.

Weber, Max. 2004. *The Vocation Lectures: Science as a Vocation, Politics as a Vocation.* London: Hackett.

1 The politics of public reason*

Sheila Jasanoff

Politicizing science

Science and technology are commonly taken as drivers of social change. Less visibly but quite centrally, as this book argues, they are also crucially important objects and instruments of politics.[1] What happens in the course of knowledge production, and still more plainly in the translation of knowledge into technologies, affects the kinds of lives we lead, the relationships we form, and, increasingly, how we perceive ourselves and what entitlements we therefore claim. All of the traditional categories of social organization – race, class, gender, nationality, ethnicity, economic and professional status, occupation and family – have been profoundly reshaped in modernity's long march through the scientific, industrial and high-tech revolutions. Whether we see ourselves as enlightened, globalized, networked or knowledge societies, those era-defining terms themselves reflect epistemic and social configurations that would not have been possible without fundamental changes in science and technology. Hence, science and technology are fitting though strangely neglected subjects for political analysis.

The project of politicizing our understandings of science and technology holds formidable difficulties. One measure of science's extraordinary institutional successes over the past four centuries is that intellectuals have by and large fallen out of the habit of thinking that there is anything political about science, either as a domain where politics happens or as a subject matter that is influenced – except in the most mundane and corrupting sense – by politics. Almost by definition, science is the sphere of incontestable knowledge, a space that both is and should be immune to politics because it is simply about being truthful to nature. We get to science, conventional wisdom holds, precisely when we have shorn away values, conflicts, passions, desire, emotions, interests; in short, all those things that make up the stuff of politics. It follows that to politicize science seems in principle a forbidden act; it sounds suspiciously like settling the truth by popular vote, by economic power or by *diktat*. It is how to keep politics out of science, often described as preserving scientific integrity, that has preoccupied the politics of knowledge making for many decades.

When it comes to technology, the erasure of politics has been somewhat less complete, but even then political analysis largely limits itself to looking at the ways in which technological systems affect people's lives.[2] Politics, on this view, comes into play principally when technology is implicated in limiting or enhancing life, liberty or the pursuit of happiness. Innovation proceeds, so it is believed, through the natural operations of human ingenuity and the market. What needs reflection, and possible political control, is only technology's potentially harmful impacts or, in standard apolitical parlance, its unintended consequences.

If persistent depoliticization is one part of the problem for a deeper politics of knowledge, then a further difficulty is the lack of developed theoretical frameworks in which to ask, or answer, penetrating questions. What should political analysis concentrate on when the subject matter is something so abstract and seemingly bloodless as knowledge? And, even if one grants that politics and epistemology are somehow related, how should analysts deal with the fact that the discourses in which we speak of politics constitute a knowledge domain that itself needs critical reflection? Is such reflexivity possible or will it paralyze normative as well as epistemic analysis? Little in the classical works of political philosophy from Thomas Hobbes to Juergen Habermas invites today's scholars or citizens to consider how the fundamental categories of political thought – such as power, representation, and democracy itself – have been modified or recast in light of the far-reaching transformations wrought by science and technology.

A full-blown political analysis of science needs to ask not only how to expose the values inadvertently locked up in the spheres of science and technology, but also the symmetrical question of how unexamined assumptions about democracy get reified through the intrusions of science and technology into the public sphere. What kinds of democracies are possible, imagined or actualized in technoscientific societies? Why do ways of steering science and engaging publics differ even among closely similar societies? Are there good and less good ways of relating politics and science? Put differently, what can democratic theory gain if we perceive science and technology not only as sites but as active agents of political production, enabling different forms of democracy to come into being and to perpetuate themselves with little further self-questioning? Those are the questions that this chapter chiefly addresses.

Two principal arguments run through the chapter. First, divergent accounts of the right relationship between science and democracy reflect historically and culturally situated conceptions of how science governs itself. Second, as articulated into political practice, these theories underwrite radically different constructions of the human subject as political actor and agent of democracy. I begin by sketching two diametrically opposed strategies for resolving the tensions between science and politics: separation through firm demarcation; and integration through broad public participation. I then show how theories of scientific knowledge-making have underwritten particular modes of demarcation and participation; and how the incorporation of these theoretical positions into contemporary political practices affects the production of public reason. I conclude by showing that the demarcationist and participatory approaches correspond to fundamentally different constructions of the human subject as a knowing political agent. These variations in the theories and practices of public reason need to be unpacked and assessed if we are to approach the democratization of science and technology as a meaningful project.

Science and democracy: from demarcation to engagement

Democratic theory has not been completely oblivious to the links between science and politics. In particular, the paradox of creating influential but apolitical preserves within democracies has been a source of continuing vexation for theorists as well as concerned citizens. As societies have come to depend more heavily on science and technology, the need to look critically at the relationship between expertise and democratic values has been widely acknowledged. If elected officials rely on unelected experts to govern, how can public decisions remain accountable and subject to democratic control? Worries have grown, especially in European political thought, about the power of technical rationality when joined with instruments of state to organize and orchestrate life, discipline it, condition its possibilities for expression and individuation, and at the limit to obliterate human-ness or make it not worth cherishing. Both political theory and political practice have wrestled with such issues, oscillating between two major prescriptions for striking the right balance between science, technology and politics: first, demarcation, to define and keep well apart the spheres of facts and values; second, participation, to ensure that people's voices are heard and acknowledged deep within the perimeters of technical decision making.

Historically, demarcation came first, reflecting concerns about the dominance of expertise in public life, and the resulting risk of hyper-specialization and narrowing of vision at the expense of wisdom and sound judgment. Critics believed that these problems could best be solved by restricting scientific experts to clearly defined technical and advisory roles, "on tap rather than on top" in a neat turn of phrase attributed to Winston Churchill.[3] By the 1970s, however, both law and policy reframed the democracy problem as having less to do with experts usurping the role of elected representatives and more to do with experts' lack of accountability to lay publics. Openness became the new watchword, and procedural creativity the chief means to implement it: from transparency rules for advisory committee meetings to freedom of information and expanded opportunities for publics to question administrative decisions.[4]

Concerns about the power of experts – in relation to public officials as well as citizens – were premised on an inarticulate mix of epistemic and normative assumptions. On the epistemic front, the demarcationists tended to accept that experts do know best, at least about the subject matter within their control. Their doubts centered on classical principles of democratic delegation, in which elections and politically accountable appointments, not superior knowledge, form the only legitimate basis for authority. Problematic too was the felt difference between expert knowledge and the knowledge needed to govern well, in short, the difference between specialist and sage. Relegating technical expertise to a place apart seemed to be the only practical answer. Insulated from the messiness of politics, science would be free, in the words of Don K. Price, a founder of Harvard University's John F. Kennedy School of Government, to "speak truth to power." If politics ignored the voice of disinterested truth, it would do so at the risk of its own credibility and effectiveness. This is the logic that, forty years later, propelled science writer Chris Mooney's 2005 book, *The Republican War on Science*,[5] to best sellerdom. Mooney lambasted the Bush administration to great effect precisely for having turned its back on well-recognized scientific truths.

Advocates of participation, by contrast, see the promise of clean demarcation as grounded in false or misleading presumptions. Expert knowledge claims, according to proponents of participatory democracy, often conceal undisclosed and undiscussed assumptions, opaque, untested, and inevitably value-laden. Expertise, in other words, tends to gather under its depoliticizing umbrella matters that are properly political. Citizen participation then is the antidote to a kind of overbroad delegation that allows experts, operating beneath layers of epistemic claimsmanship, to seize control of democracy's normative agendas. By opening up expert judgment to lay review, this line of criticism seeks to give power back to people – restoring to citizens their rightful place in the politics of expertise.

For many decades, the two logics for balancing science and politics coexisted in relative harmony, though they were organized around different understandings of the politics of knowledge and called for very different strategies of implementation. By the turn of the twenty-first century, however, signs of friction set in. For old-style demarcationists, charity toward lay opinion seemed to have gone too far. Increased participation raised fears of a new populism, resistant to modernization, technological progress, and the benefits of globalization. Episodes of seemingly irrational behaviour – such as opposition to the measles–mumps–rubella (MMR) vaccine as a possible cause of autism in the United Kingdom,[6] the European resistance to genetically modified (GM) crops, South African president Thabo Mbeki's ill-conceived rejection of the viral theory of HIV-AIDS, and the US administration's failure to react to climate change under President George W. Bush – appeared to confirm fears that public ignorance was overwhelming reason. A 2007 letter to *Nature* put the point crisply:

> SIR – Last night I had a nightmare. In my dream, all the recommendations made by Pierre-Benoit Joly and Arie Rip in their Essay "A timely harvest" (*Nature* 450, 174; 2007) became a reality here in the United States. The public were consulted and actively engaged in practical scientific matters.
>
> I dreamed that the dos and don'ts of science and research were dictated democratically by the American public, of whom 73% believe in miracles, 68% in angels, 61% in the devil and 70% in the survival of the soul after death. In my dream, this majority dictated through vigorous "public engagement" that science should deal with virgin birth, the thermodynamics of hell, the aerodynamics of angel wings, and the physiology and haematology of resurrection.[7]

In response to such fears, some tried to formulate new ground rules for public involvement, so that science and technology policy would stay out of the clutches of the ignorant, the fearful and the politically opportunistic. From this protectionist vantage point, the problem of democracy was in effect up-ended: not how to make ruling technical elites more accountable to the governed, the chief concern of many twentieth-century democracy theorists, but how to keep the unruly *demos* (and its unreasoning representatives) out of places where they had no right to be.[8] The new defenders of expertise argued that publics must meet threshold criteria of knowledgability before they could interact with experts, let alone overrule them.

Defenders of democratic participation, among whom I count myself, view this retreat to a given-in-advance demarcation of expert from non-expert spaces as intellectually and normatively untenable.[9] Such demarcation, to begin with, rests on prior political choices, as documented by a generation of work in science and technology studies.[10] Values inevitably get sequestered within the domains labeled as science or expertise and are shut off to deliberation unless opportunities exist to look behind the labels. Public opposition may signal not unreason but the failure to take relevant values sufficiently into account. One reaction against the new demarcationists has therefore been to push the logic of public participation still more aggressively: "upstream," into processes of research policy and technological design. The call then is for "public engagement," a variety of proactive measures to involve people in shaping the purposes of research and the design of technologies. Britain's nationwide public consultation on agricultural biotechnology, the 2003 *GM Nation?*, was perhaps the best known European experiment in engaging publics so as to transcend the limits of end-of-pipe, product-focused, risk-based regulation.

Producing knowledge, reproducing politics

Both the reactionary and the radical solutions to the problem of participation converge in one respect: both focus mainly on the production side of science and technology, asking who should govern science's infinite potential to engender novelty. For today's technocrats as for yesterday's, the question "who knows best?" looms largest, fed by fears that ceding power to the non-knowledgable could lead to unreasoning restrictions on research and innovation. For the resurgent democrats, the key question is about people's right to determine the directions of social and material progress, a goal that demands wider public engagement. From both perspectives, though, science and technology are conceived more as things shaped by politics than as players shaping the essence of the political.

That second direction of the arrow can no longer be ignored. If, following W.B. Gallie, we take democracy to be "essentially contested,"[11] then the focus of political thought necessarily shifts from "how do we achieve this thing called democracy?" to "what is this thing called democracy that we hope to achieve anyway?" This move becomes highly consequential if we recognize that science and technology are agents in the co-production and continual re-enactment of social and political orders.[12] On this view, science and technology are not only epistemic and material but also normative processes: they are part of the dynamic of building integrated understandings of how the world is and how it should function, with combined support from expert knowledge, material innovation, and political power. In determining how to govern the production and uses of scientific knowledge, we necessarily also reflect on and reaffirm particular conceptions of political order. It is not enough, then, simply to ask for more participation. One must also ask correlative questions about the theories of democracy and rationality that underpin particular approaches to broadening the public's role.

Below, I develop a view of science and technology as agents of political production. Steering between the poles of defensive demarcation and interventionist engagement, both of which take some sort of preordained separation between science and the political process for granted, I show that speaking about science in democracies is political all the way through. What societies choose to protect, what authority they invoke, and what they challenge under the rubric of "science" all relate in complex ways to the things they wish to protect, invoke, or challenge under the rubric of "politics." Making sense of the politics of knowledge thus requires, in part, deeper reflection on the contested nature of contemporary democracy, most particularly those aspects of democratic governance that rationalize or naturalize the exercise of power. My project, then, is as much to open up the epistemic presumptions underlying diverse democratic settlements as it is to problematize the unseen or hidden values entailed in the production of scientific knowledge or its technological manifestations. I show that the commitment to public reason in modern societies rests in tacit ways on divergent, at times nation-specific understandings of politics, citizenship, expertise, and of science itself. To this end, I tease apart the disparate presumptions embedded in three processes that are central to the contemporary politics of reason: first, bounding science off from politics; second, the delivery of expert advice; and third, the performance of reason in the public square.

The politics of boundaries

In an influential essay, the philosopher and sociologist of science Bruno Latour claimed that a defining feature of modernity is the separation of hybrid worldly phenomena into clearly demarcated domains of nature and culture.[13] This was no mean achievement when everything that exists can be seen as criss-crossing that quasi-constitutional divide. Latour was eloquent in arguing that cloned sheep and ozone holes, cattle diseases and streetcars, scientific theories and mundane door keys, all come into being and are sustained through dense interconnections among humans and non-humans, animate and inanimate, society and nature. Metaphysics aside, the separation that Latour postulated had immense consequences for modern democracies. The notion that science exists as an autonomous form of life, constrained only by its truthfulness to nature, has permitted science to chart its own research trajectories, legitimate controversial policies, and authorize political institutions to undertake large technoscientific projects, from nuclear power to mapping and sequencing the human genome, that reach ever more consequentially into people's lives.

Today, the commitment to science's autonomy has if anything regained strength among liberal intellectuals, prompted in part by seemingly unfounded public resistance such as the episodes described above. But science's claimed independence from political control is itself built through social means. A key insight imported from sociology into science and technology studies is that what counts as science in particular settings is often contested, precisely because so many social and political consequences hang on that designation; the line separating science from pseudo-science, religion, politics or mere subjective belief is discernible as such only after contestation ceases, following intensive boundary work. The boundary between science and politics is foundational, but drawing that particular line in the sand may involve substantially different conceptions of how science works.

Thomas Kuhn, the twentieth century's most cited philosopher of science and the figure most often credited with fathering the social studies of science, saw science as a social activity, but the dynamics of knowledge-production were, in his account, largely internal to science itself. In this respect, Kuhn performed a significant though tacit form of boundary work. Scientific revolutions happen, he famously observed, when a theoretical paradigm can no longer contain all the contradictions built up within it through what he called "normal science." Causes of revolutionary change, in other words, are to be found inside, not outside, science. That inwardness helps explain why Ian Hacking, an eminent philosopher of science of a later era, briskly dismissed Kuhn as a social analyst:

> Kuhn said little about the social. More than once he insisted that he himself was an internalist historian of science, concerned with the interplay between ideas, not the interactions of people. His masterpiece, ever fresh, is now over thirty-five years old – truly the work of a previous generation.... Yet for all that Kuhn emphasized a disciplinary matrix of one hundred or so researchers, or the role of exemplars in science teaching, imitation, and practice, he had virtually nothing to say about social interaction.[14]

Historians since Kuhn have argued that scientific revolutions bring along with them social and political rearrangements that are little short of revolutionary. Radical shifts in scientific and social order are, in this view, coproduced. This in effect is the meta-argument advanced in Steven Shapin and Simon Schaffer's seminal work on the history of the scientific revolution in seventeenth-century England.[15] Yet Kuhn's vision of a science answering mainly to its own inner drives still retains power,[16] and it still helps authorize science's image as a realm apart from politics.

The renowned chemist and philosopher of science Michael Polanyi acknowledged the possibility of politics within science, as well as the subordinations and dependencies that can arise between scientific and political institutions. He concluded nevertheless that science could remain free from external controls because it is adequately self-regulating. Scientists conduct their work, Polanyi argued, divorced from thoughts about the impacts of their work; science therefore bows to no higher authority and functions as a republic in and of itself. From this followed Polanyi's famous dictum: "Any attempt at guiding scientific research towards a purpose other than its own is an attempt to deflect it from the advancement of science."[17] To illustrate, he told a self-deprecating story:

> An example will show what I mean by this impossibility. In January 1945, Lord [Bertrand] Russell and I were together on the BBC *Brains Trust*. We were asked about the possible technical uses of Einstein's theory of relativity, and neither of us could think of any. This was forty years after the publication of the theory and fifty years after the inception by Einstein of the work which led to its discovery. It was fifty-eight years after the Michelson-Morley experiment. But, actually, the technical application of relativity, which neither Russell nor I could think of, was to be revealed within a few months by the explosion of the first atomic bomb. For the energy of the explosion was released at the expense of mass in accordance with the relativistic equation e = mc2 an equation which was soon to be found splashed over the cover of *Time* magazine, as a token of its supreme practical importance.

Polanyi meant his audience to chuckle at his own and Russell's naiveté, but he too provided important fodder for strict demarcation by insisting that pure science is honest and self-disciplining because it has no social value, or at least none that scientists see as relevant. The bomb may be political on Polanyi's account, but physics is not. It is only when science morphs into technology that values and politics come into play.

Robert K. Merton, America's pre-eminent mid-century sociologist of science, rested his case for scientific autonomy on a still different reading of what keeps scientists honest. In his famous 1942 essay, "The Normative Structure of Science," Merton argued that science remains true to its institutional ethos of truth making, and succeeds best at it, when it imbibes, and reproduces, ethical commitments that are characteristic of democratic and capitalist societies. The commitments, or norms, that Merton identified as core values of science were: *universalism*, the testing of truth claims according to impersonal, non-local criteria; *communism* (or *communalism*), the common ownership of scientific knowledge; *disinterestedness*, the pursuit of knowledge detached from economic or political uses; and *organized skepticism*, the questioning of all forms of vested authority, including the religious or sacred. While Merton recognized cases of science thriving in anti-democratic or, more accurately, pre-democratic societies, modern science he opined should be free from state supervision because it adheres to the same norms of trust-building as democracy itself.

Boundary work of the kinds conducted by Kuhn, Polanyi, and Merton is of interest today precisely because it is more than an academic abstraction. It is grounded in observations of practice, which it seeks to distil into principled generalizations. Politics, more than science, is the place where such boundary-making practices can be observed most clearly: political actors have long sought to demarcate science from politics while drawing on scientific authority to sustain their own legitimacy. Those political demarcations rest in turn on more or less explicit theories of how science works. Thus, we find echoes of philosophers and sociologists of knowledge circulating through the political work of contemporary democracies – though unrecognized as such and without attribution. In the United States, for example, Merton's ideas routinely surface in political speech and action. In a 1990 address to the National Academy of Sciences, President George H.W. Bush alluded to the convergence between science and democracy in terms reminiscent of Merton's a half-century before:

> It's no accident that many of the individuals at the center of today's worldwide political revolutions share a vision of the future based on personal freedom, openness, and freedom of inquiry. These values are shared by our political system and by science alike.[18]

Even more strikingly, the norm of organized skepticism celebrated by Merton finds continual rearticulation in a long-standing and peculiarly American concern with the peer review of policy-relevant science. Congress, regulatory agencies, the Office of Management and Budget, industry representatives, and scientists themselves have been drawn into intense, recurrent debates over the appropriate form, function, and control of regulatory peer review. In such debates, peer review is advanced as a necessary and sufficient guarantor of scientific objectivity. As I have argued elsewhere,[19] this strategy displaces concerns about the objectivity of science onto a second-order politics of selecting reviewers who will sign off on particular readings of evidence in contested domains. In this way, the image of science as pure and non-manipulable retains power as a potent rhetorical resource while people continue to fight about how scientific knowledge should be validated.

By contrast, conflicts about the design and conduct of peer review have been almost non-existent in other Western nations, even when governments confront perceived breakdowns in the relations between science and politics. It is as if Europeans live in a world in which Mertonian sociology has taken the back seat to other theoretical justifications for boundary work between science and politics. In Britain, consistently with Polanyi's views, political controversies implicating science often revolve around debates about scientists' virtue and competence; so long as scientists are seen as adhering to communal norms, the authority of science is deemed to be in safe hands.[20] In Germany, where institutional processes draw less clear distinctions between scientific and political consensus-building, US-style public controversies over technical facts and evidence are virtually unknown. As we see in more detail below, divergences in the theorization of science have become especially salient in disputes about the legitimacy and limits of expert advice.

The politics of expert advice

Scientific facts do not arrive from nowhere, by divine intent or natural law. It takes human work to make nature speak, and the representations resulting from such work are partial, provisional, possibly clouded images of whatever reality lies behind. More important, "science" as such never directly speaks to power. Normal science, when operating in tune with its Kuhnian rhythms, seldom holds answers to the questions of urgent interest to policy makers. On almost every issue with a technical component that matters, it is not *science* that states rely on so much as expert judgment. And no one has ever claimed that experts "speak truth to power." What, then, are the assumptions that lead democratic societies to develop, and at times to disavow, trust in expert advice on delicate and contested matters of state? Here again we find that reliance is underpinned by divergent sociologies and philosophies of knowledge carrying persistent consequences for the ways in which a government's advisory business is conducted.

In Britain, loss of trust in expertise became a political leitmotiv at the turn of the twenty-first century. The "mad cow" crisis, sparked by the discovery that a fatal brain disease had jumped the species barrier between cattle and humans contrary to expert predictions, followed by widespread public rejection of GM crops and MMR vaccine, led to fears of a more pervasive corrosion of trust. A 2000 House of Lords Select Committee report characterized the relations between science and society as being in a "crisis of confidence."[21] The Labour government's attempts to restore trust led to significant changes in British institutional practices of soliciting expert advice on issues ranging from food safety to embryonic stem cell research. Yet a controversy surrounding the dismissal of a senior adviser on pharmaceutical regulation in late 2009 revealed that the discontent identified by the House of Lords report was still simmering beneath the surface of workaday British politics. Responses from both science and government revealed a theory of the right relations between knowledge and politics that was more consistent with Polanyi's demarcationist model than with Merton's.

Professor David Nutt, chief of the UK government's Advisory Committee on the Misuse of Drugs (ACMD), was abruptly fired in October 2009. That action followed an ACMD review of the medical evidence supporting the reclassification of cannabis (marijuana) from the less serious Class C to the more serious Class B status. The committee concluded in 2008 that there was little risk of cannabis causing serious mental illness, as had been alleged to justify the more stringent classification. Nutt himself publicly declared that he considered the risk of psychotic illness from cannabis to be minimal. His subsequent dismissal on the ground that the government no longer had confidence in his advice caused a media firestorm, embarrassing politicians and enraging many scientists. To calm passions, the government issued a set of guidelines designed to improve working relations with its scientific advisers and to restore public trust.

In all the noise surrounding Nutt's dismissal, what stands out is the rhetorical emphasis on the independence of science. On December 15, 2009, about six weeks after Nutt left office, the Department for Business Innovation and Skills issued for consultation a set of high-level principles of scientific advice, designed "to ensure effective engagement between the Government and those who provide independent science and engineering advice."[22] Commenting on the action, the minister, Lord Drayson, emphasized the centrality of academic freedom, as well as the mutual responsibilities of science and government. "Independence" receives a separate section in the draft principles, which include a provision that scientific advisers should be free from political interference with their work. Government's obligations are summarized primarily under the heading of "Transparency and openness" in the timing, explanation and publication of responses to advice. Peer review makes no appearance in these principles, nor does any mention of a role for citizens or publics at large. In the allusion to academic freedom and the rhetoric of independence, one hears echoes of Polanyi's "republic of science," a self-regulating polity that can be let alone when functioning according to its own norms. The logic underlying the proposed restoration of trust is strictly demarcationist: let science be science; let politicians heed independent advice. Little in this back-to-basics proposal to suggest that Lord Drayson or his department had heard or noticed the demands for upstream engagement and more thorough democratization of science that were significant markers of late twentieth-century British politics.

The politics of performance

The politics of science and government plays out not only in the formal pronouncements of national leaders and the routines of advisory committees, but in responses to unplanned events that call forth unrehearsed shows of official knowledge and competence. In such episodes, public reason giving operates as a kind of theater, a performance which, when effective, reasserts the state's right to wield power on the basis of its proclaimed logics and knowledges.[23] Governmental attempts to restore reason, often in the aftermath of breakdowns and failures, enact particular understandings of democracy: who is the audience, what sorts of demonstrations does it need, and so forth. When public reason fails, it is often because lead actors did not abide by the cultural scripts and codes that define what I have termed civic epistemologies – that is, "shared understandings about what credible claims should look like and how they ought to be articulated, represented, and defended."[24] Such failures look much like failed experiments in garnering public trust.

On March 20, 1996, Britain became the unwitting host to just such an experiment in democracy. That was the day when Stephen Dorrell, Secretary of State for Health, announced to the House of Commons that ten cases of a new, "variant" form of Creutzfeldt–Jakob disease (vCJD), most likely linked to the ten-year-old "mad cow" epidemic, had been diagnosed in human patients. The announcement came as a shock to a nation that had been told for almost a decade that the chance of disease transmission from cows to humans was virtually nil. Issuing from government ministers and their expert advisers, the refrain that "British beef is safe to eat" had circulated through the media with boring consistency. Now, suddenly, not only was that reassurance shown to have rested on hollow ground, but it seemed that the government had been so unsure of its claims that it had secretly collected data to test its own hypothesis – data that eventually disproved the hypothesis and grew too compelling to hide from the public.

The sequel is now well known: widespread flight from beef-eating in the immediate aftermath of the government's disclosure; long-lingering suspicion and distrust of the government's expert advice; numerous formal institutional modifications designed to undo the damage, including the establishment of the Food Standards Agency in 2000; and later public rejection of GM crops and foods. There was, however, a telling piece of public theater that occurred while the "mad cow" crisis was brewing in 1990, intended as reassurance, but in its long-lingering effects a fiasco for the state.[25] That episode exemplifies the importance of informal performances as supplements to official knowledge claims, and the crucial role of civic epistemology, an often overlooked element in the politics of public reason.

The lead performer in that earlier drama was John Gummer, then Minister of Agriculture with frontline responsibility for assuring the safety of the food supply. In an ill-fated move, Gummer held a news conference in his Suffolk constituency, during which he attempted to feed his four-year-old daughter Cordelia a beefburger. What Gummer hoped to demonstrate with this public display was not simply a father performing a basic ritual of parenting; it was also a performance of the British state in full command of its expert knowledge, prepared to act as *parens patriae* (parent of the nation) for all its citizens. But the public, helped by the very media the minister sought to enlist, famously rejected Mr. Gummer's demonstration. The pictures he commissioned from the media gave rise to readings, in words and images, that lampooned the minister's intended message. Contemporary newspaper and television accounts poked fun at Gummer's apparent desperation in trying to stage a persuasive public drama; Cordelia's reluctance to eat the beefburger became a lasting metaphor for public rejection of governmental insincerity.

Events in 1996 brought Gummer's gaffe back into the limelight. For weeks, British public servants were asked to bear personal witness that they considered British beef to be safe. With the policy apparatus seemingly in stalemate,[26] public officials were badgered for personal declarations that they had not stopped eating homegrown beef. John Pattison, the government's chief scientific adviser on variant CJD, admitted that his grandson was not allowed to eat beef.[27] Perhaps less candidly, a dozen or so cabinet ministers told the *Independent* that beef was still allowed on their family dinner tables.[28] Against this backdrop, a brilliant satirical image produced by the political cartoonist Gerald Scarfe resurrected Gummer's performance as a metaphor for state hypocrisy.[29] Above the caption, "What have we been fed?," Scarfe depicted a black-clad, male figure, forcibly stuffing a hamburger into a child's mouth, her head roughly held back (as it was not in real life). The abusive cartoon parent was a far cry from Gummer, the well-intentioned conservative Christian, son of a clergyman, who had innocently used his daughter as a surrogate for the public he was committed to serving. His abortive gesture, however, brought to light a silent shift in the democratic gaze that had occurred without the British state seeming to take note of it, placing new interpretive authority in the eye of the beholding public and engaging its critical faculties in unforeseen ways.[30]

A similar performative moment occurred in the United States in the wake of Hurricane Katrina, the devastating storm that hit the city of New Orleans in 2005 and contributed to the unraveling of trust in the George W. Bush administration. As in Britain's "mad cow" crisis, the hurricane created conditions that put high political authorities on the spot to offer explanation, reassurance, and redress. To be sure, the circumstances were entirely different: not a creeping, fatal disease, caused by a poorly understood agent, liable to strike with uncertain probability and no notice, but a localized storm, with a well-defined beginning and end, bringing tangible chaos, destruction, and death in its wake. Both events, however, required the state to engage in visible displays of public reason. After Katrina, as after the disclosure of variant CJD in humans, the government had to field questions about why it had not known enough to take proper precautions, and why it had not acted to mitigate damage more effectively once disaster struck.

Katrina, too, produced a figure of fun comparable in some ways to John Gummer. This was Michael D. Brown, Under-Secretary of Emergency Preparedness and Response, and Director of the Federal Emergency Management Agency (FEMA). At a moment when a fabled US city was drowning on national television, and criticism was mounting about the federal government's fumbling and inadequate response, President Bush threw his support behind the beleaguered FEMA chief in words that resonated, for sheer misguided timing, much like Gummer's ill-conceived televised stunt: "Brownie, you're doing a heck of a job!" In the US as in the British case, a leader's performance flew in the face of things the public could see and judge for itself, eliciting disbelief and deepening the loss of trust. Years later, Bush's words still drew snickers and prompted administration loyalists to offer defensive interpretations, showing how badly those few words had misfired.[31]

From the standpoint of the politics of knowledge, however, something importantly different was at stake in the US case. The state's mistake in Britain had been to treat the public as ignorant children, incapable of accepting that scientific predictions might be clothed in a penumbra of irreducible uncertainty, that prediction is always probabilistic, and that there is no such thing as zero risk. Indeed, in testimony to a European Parliamentary committee, the late Sir Richard Southwood, chair of the government's first Working Party on BSE, acknowledged this failing:

> I believe that the scientists must try to indicate the probabilities of various outcomes. It is easy to say the hypothesis has not been tested and therefore there may be a great risk or none at all. This is a way of ensuring that one is not wrong, but in my personal view such an evasion of providing guidance is a dereliction of one's duty, as a scientist, to society.[32]

This comment reflects Southwood's own learning process, since his committee was publicly faulted for failing to disclose the uncertainties of "mad cow" disease in timely fashion. Here he acknowledged that it would be better for science to represent multiple possible outcomes, with probabilities attached, than to claim absolute certainty.

Yet, poignant as his words seem, they home in on a solution that still construes relevant epistemic uncertainty as belonging in the care of scientists – not experts who *judge*, but scientists who *know*, and who, as Southwood asserts, have a duty to communicate their knowledge accurately to the people. It thus becomes a moral mistake (a "dereliction of duty") on the part of scientists not to offer guidance on alternative probabilistic scenarios. But as I have argued elsewhere, the Southwood Working Party's original failure to acknowledge the risk of cross-species disease transmission was not at all a matter of dereliction. The committee did its duty, but its performance was tied up with aspects of British civic epistemology: those deeper cultural tendencies in Britain to accept Polanyi-style characterizations of science as an autonomous republic; to adopt the empiricist's fondness for assuming facts to be clean and neat; to privilege the evidence of the senses and to downplay uncertainty; and, correlatively, for scientists and officials to underestimate the public's ability to deal maturely with endemic uncertainty, the absence of any definitive answers to complex technical questions.[33]

Civic epistemology played out somewhat differently in the aftermath of Katrina. Here the issue was not the intellectual failure of institutional science but the ineffectual behavior of experts who are supposed to act as translation agents between science and society. The facts that resonated most with the public about FEMA chief "Brownie" were his incompetence, reflected in his lack of prior expertise in emergency management, and his political connections to the Republican Party, which had apparently appointed him to a post whose functions he was unqualified to carry out. What the American crisis brought to light was the gap between the culturally valued professionalism of high-ranking experts (as competent translators of objective science) and the distorting reality of political influence. In the British case, by contrast, no one impugned the personal motives or competence of individuals entrusted with official responsibilities throughout the long "mad cow" crisis. Rather, official inquiry stressed the failure of those responsible, including the Southwood Working Party, to see facts that should have been plain, and to take actions that properly reflected the ascertainable uncertainties.[34]

The politics of reception: knowledgable citizens

I have argued thus far that theories of how science, technology, and expertise relate to politics vary, and that the practices of democratic politics instantiate these divergent theoretical positions in subtle, culturally inflected ways. The theories of analysts such as Kuhn or Merton or Polanyi can be reconsidered in this light not as objective accounts of how science works but as historically and culturally situated artifacts of particular social practices of boundary drawing, expert validation and political performance. In this concluding section, I turn the lens away from constructions of science and expertise in relation to politics to the construction of democracy in relation to science. I describe two persistent, mutually contradictory movements in the characterization of citizens with respect to public knowledge and public reason. Crystallizing over the past half-century, these competing conceptions of citizenship reflect a deeper struggle between a depoliticized, universalizing image of science and one that is always already engaged in the dynamics of democratic politics. That struggle underscores this chapter's emphasis on the need for democratic theory to incorporate not merely the production but also the reception of scientific claims more actively into accounts of the politics of knowledge.

The practices of depoliticization that bound science off as an apolitical space often go hand in hand with the construction of lay publics as scientifically illiterate, and hence unfit to participate fully in governing societies in which scientific knowledge matters. That characterization of the layperson, which underpins demarcationist arguments, has taken shape under a concept that looks democratic on its face but has turned out, in application, to be enormously disempowering of citizens. This is the notion of the "public understanding of science," abbreviated as PUS or, with the addition of technology, PUST. Since the mid-1970s, many Western nations have sought to measure how much or how well their publics understand science, ostensibly so that scientific performance, and by extension the lay citizen's capacity to function effectively in knowledge societies, can be improved through public policy. These measurements are deemed essential for democracy because – without basic scientific knowledge – no citizen of the modern world could think rationally or endorse the reasoned conclusions of government decision makers. Knowing science according to criteria that national policy makers accept as dispositive operates in this respect as a necessary condition of citizenship in many contemporary democracies.

Sadly for this view of democracy, the report cards of science literacy have proved disappointing. PUS measurements typically demonstrate that publics do not know many of the simplest facts of science. Even the statement that the earth goes round the sun was not recognized as true by some 30 percent of the adult US population surveyed by the National Science Foundation (NSF) in 2006, and less than half accepted the theory of evolution as true.[35] NSF indicators also revealed distressingly high levels of belief in paranormal phenomena ranging from extra-sensory perception to haunted houses, clairvoyance, and even the existence of witches. Added to this is a persistent belief in astrology, which some 30 percent of Americans and apparently 50 percent of Europeans consider scientific. For PUS advocates, not only science but democracy is at risk when large segments of the public seem to have retreated so completely from the legacy of the Enlightenment. Demonstrated gaps in knowledge and understanding are seen as threatening the very possibility of a shared public sphere, leading instead to beliefs in pseudo-science and alternative medicine, undermining support for scientific research, and (especially in the United States) offering aid and comfort to religious fundamentalists, proponents of creationism, and holders of fringe beliefs.[36]

But what is the political meaning of putting such elementary factual questions to adult national publics? What do the survey designers hope to accomplish, and what use are measurements of public ignorance in societies whose survival depends on going about well with the products of science and technology? As critics have repeatedly observed, PUS surveys do not merely test a respondent's understanding of science: they simultaneously construct the respondent as a particular kind of knower, or more accurately a *non*-knower.[37] A person is seen as ignorant, irrational, and presumptively incompetent to participate in governance if he or she does not know a preordained set of facts or holds beliefs that are deemed inconsistent with scientific rationality. Yet, as even the first-generation demarcationists recognized, the knowledge that contemporary societies need for governance is never so neatly programmed, nor so unequally distributed among experts, policy makers, and lay people, as suggested by these snapshots of public misunderstandings. Democracies do function well enough on the whole, despite their flaws, failings and inadequacies; and they do best when striving for inclusivity of beliefs, without which they would not be democratic. Are there then other models of citizenship to offset against the ignorant masses, incapable of self-governance, revealed by the PUS surveys?

Countering the hand-wringing over scientific illiteracy, a parallel notion of citizens as knowledgable, or more accurately knowledge-*able*, has indeed been developing in Western administrative law and practice for more than half a century. That evolution has given citizens increasing power to question the technical reasoning of governments, to offer counterarguments and countervailing expertise, and to challenge public reasoning that appears unsubstantiated, unbalanced or politically motivated. These "knowledge rights" – as we may call them – originate from varied sources in contemporary legal and policy systems, and their spread has been unmistakable, if patchy and gradual, through all contemporary democracies. Unlike PUS surveys, these developments presume competence and skill as baseline human endowments that citizens do not have to prove or demonstrate through test-taking.

The rights of citizenship in today's knowledge societies, articulated in laws such as those indicated in parentheses, include the following:

- Right to know:
  - of exposure to risks (freedom of information laws)
  - for informed consumption (consumer protection laws)
  - for fairness in litigation (discovery rules for litigants)

- Right to give informed consent (rules for protecting medical patients and human subjects in research)
- Right to demand reasons (administrative procedure laws)
- Right to participate and offer expertise (rules for consultation and participation)
- Right to challenge irrational decisions (environmental, health, and safety laws)
- Right to appeal adverse decisions (laws granting access to courts).

Taken together, this bundle of rights has radically transformed the political fabric of knowledge societies. They are as critical for the exercise of citizenship as are the constitutional guarantees of personal liberty, and they envisage a significantly more proactive role for publics in the politics of knowledge than the deskilling and exclusionary PUS model.

The model of the knowledge-able citizen emerging from administrative law and practice presumes that effective democratic participation is based on good reasoning, not selective factual knowledge, on informed skepticism, not blind trust, and on dynamic, situated knowledge acquisition, not a static repertoire of uncontested facts. Knowledge-able publics are deemed capable of holding their rulers to high standards of reason provided they have procedural opportunities to do so. Crucially, this model presumes that publics can be trusted to exert their judgment well if they have the means to inform themselves as need arises. Accordingly, the knowledge rights approach stresses openness, transparency, and interaction as key values. It presumes that relevant facts will be acquired as needed by publics who have an interest in getting involved in the politics of reason.

Some of the dramatic failures of expert legitimation discussed above can be understood in the light of this analysis as results of the unresolved tension between the PUS and knowledgable citizen models. The "mad cow" crisis in Britain looks in retrospect like a consequence of democratic exclusion based on a patronizing assessment of the public's incapacity to understand uncertainty. Cocooned in its cozy world of industry supporters and supportive experts, the now defunct UK Ministry of Agriculture, Fisheries and Food failed to see that trust and legitimacy flow from continual interaction with a knowing and skeptical public, not from decrees issued to people who cannot review for themselves the back-story that led to questionable decisions. Similarly, later skepticism over GM crops and MMR vaccine points to the inadequate engagement of the British public in productive dialogue about the purposes and uncertainties of particular technological developments. Hurricane Katrina's aftermath, by contrast, violated not only common-sense expectations of emergency response, but also American norms of neutral expertise and the professional (not political) conduct of skilled governmental responsibilities.

A practical lesson that emerges from examples like these is that, in today's science and technology-infused democracies, it makes no sense to treat the vast majority of the public as ignorant natives of foreign cultures, as the PUS model does. The law has been quicker to recognize than science that democracies are only as strong as the citizens who populate them. Patronizing or excluding citizens is not the answer to democracy's failures. What makes democracy work, whether in specialist domains of expert policy making or more generally, is continual interaction. Public understanding grows through myriad repeated encounters between people and those to whom they have entrusted responsibility: a doctor, a community policeman, a regulator, an expert adviser, or an elected political representative. That interactivity goes begging when public reason becomes artificially depoliticized, the preserve of insulated expertise. Skepticism and questioning, however uncomfortable or irritating in the short run, make better foundations for democracy than imperial opinions handed down in the expectation of acceptance.

That prescription rings an important change on the theme of public engagement. It stresses and reinforces the point that "engagement" is not simply a matter of opening the doors wide to all comers. That view trivializes the participatory process and its purposes. Rather, it is about the dynamic character of trust-building, as well as knowledge-making, in democratic societies; it is about establishing, and reperforming, the rhythms of "normal democracy" that make it possible to put one's trust in the wisdom of strangers while holding ignorance and denial at bay. If good mechanisms of engagement are in place and functioning, there will likely be no need for the multitude as a whole to demand access to expertise's innermost chambers. They will find informed delegates to represent them wisely and well.

What would have helped most in the failed technological decisions of the recent past – across all the democracies we know – is greater openness to other knowledges and other views, more opportunities for questioning and being questioned, and more spaces where alternative judgments and imaginations could have been applied to the facts at hand. This diagnosis holds true not only for BSE, MMR and GM crops in Britain, but across the Channel for the AIDS blood scandal in France and the BSE debacle in Germany, as well as across the Atlantic for the costly mistakes of 9/11 and Iraq.

The perspective I have proposed here goes beyond the usual recommendations for public engagement not only in its support for interactive processes but in its demand for more self-reflective analysis of the nature of rationality and public reason in contemporary knowledge societies. In this respect, my argument builds on the dualistic framework of co-production rather than on conventional, one-sided understandings of how knowledge and politics interrelate. For the model of the knowledge-able citizen, I have suggested, not only empowers willing citizens to participate in building democracies of reason. It also powers different possible models of democracy. Grounded in disparate institutional and social histories, and based on divergent theories of representation and voice, the knowledge rights granted to citizens in different societies reflect substantially different imaginations of good, and informed, politics. It is in teasing out those diverse imaginaries, and learning how to critique them from outside as well as within, that we enlarge our understanding of what gives public reasoning its virtue and democracy its resilience.

Notes

* A substantially less developed version of this article was delivered as the 2007 C.R. Parekh Lecture at the University of Westminster, London.

1 See, for example, Mark B. Brown, *Science in Democracy: Expertise, Institutions, and Representation* (Cambridge, MA: MIT Press, 2009); Bruno Latour, *The Politics of Nature* (Cambridge, MA: Harvard University Press, 2004); Sheila Jasanoff, *Designs on Nature: Science and Democracy in Europe and the United States* (Princeton, NJ: Princeton University Press, 2005); *The Fifth Branch: Science Advisers as Policymakers* (Cambridge, MA: Harvard University Press, 1990).

2 A line of work from Karl Marx to contemporary thinkers such as Langdon Winner has illuminated technology's capacity to constrain, and occasionally to liberate, human lives. Examples include Wiebe Bijker, *Of Bicycles Bakelites, and Bulbs: Toward a Theory of Sociotechnical Change* (Cambridge, MA: MIT Press, 1995); Lawrence Lessig, *Code and Other Laws of Cyberspace* (New York: Basic Books, 1999); and Langdon Winner, *The Whale and the Reactor* (Chicago, IL: University of Chicago Press, 1986).

3 Don K. Price, *The Scientific Estate* (Cambridge, MA: Harvard University Press, 1965); Harold J. Laski, "The Limitations of the Expert," *Harper's Monthly Magazine* 162 (1930), pp. 101–110.

4 Jasanoff, *The Fifth Branch.*

5 C. Mooney, *The Republican War on Science* (New York: Basic Books, 2005).

6 Melissa Leach and James Fairhead, *Vaccine Anxieties: Global Science, Child Health and Society* (London: Earthscan, 2007).

7 Dan Graur, "Public control could be a nightmare for researchers," *Nature* 450(20):1156 (2007).

8 An exemplary work in this genre is Harry Collins and Robert Evans, *Rethinking Expertise* (Chicago, IL: University of Chicago Press, 2007); see also Cass Sunstein, *Laws of Fear: Beyond the Precautionary Principle* (New York: Cambridge University Press, 2005).

9 See Sheila Jasanoff, "Breaking the Waves in Science Studies," *Social Studies of Science* 33(3): 389–400 (2003).

10 See in particular Thomas Gieryn, *Cultural Boundaries of Science: Credibility on the Line* (Chicago, IL: University of Chicago Press, 1999); and Jasanoff, *The Fifth Branch.*

11 W.B. Gallie, "Essentially Contested Concepts," *Proceedings of the Aristotelian Society*, 56: 167–198 (1956).

12 On co-production see Sheila Jasanoff, ed., *States of Knowledge: The Co-Production of Science and Social Order* (London: Routledge, 2004); also Bruno Latour, *We Have Never Been Modern* (Cambridge, MA: Harvard University Press, 1993); Steven Shapin and Simon Schaffer, *Leviathan and the Air-Pump* (Princeton, NJ: Princeton University Press, 1985).

13 Latour, *We Have Never Been Modern.*

14 Ian Hacking, *The Social Construction of What?* (Cambridge, MA: Harvard University Press, 1999).
15 Shapin and Schaffer, *Leviathan and the Air-Pump.*
16 See, for example, Philip Kitcher, *Science, Truth and Democracy* (New York: Oxford University Press, 2001).
17 Michael Polanyi, "The Republic of Science," *Minerva* 1: 54–73 (1962).
18 President George H.W. Bush, April 23, 1990, http://bushlibrary.tamu.edu/research/public_papers.php?id=1790&year=1990&month=4.
19 Jasanoff, *The Fifth Branch* and "Contested Boundaries in Policy-Relevant Science," *Social Studies of Science* 17(2): 195–230 (1987).
20 The public exoneration of Professor Phil Jones following the notorious "Cilmategate" controversy of 2009 offers an excellent recent illustration of this dynamic. Jones headed the University of East Anglia's Climatic Research Unit, whose e-mails were hacked by climate skeptics. Following a grueling hearing, a parliamentary committee declared that Jones had acted properly, in accordance with the norms of the climate science community. See UK House of Commons Science and Technology Committee, Select Committee Report, *The Disclosure of Climate Data from the Climatic Research Unit at the University of East Anglia* (London: The Stationery Office, 2010).
21 House of Lords Select Committee on Science and Technology. *Science and Society,* House of Lords Papers 1999–2000, 38 HL (London: The Stationery Office, 2000). The very formulation of the crisis, as one of society's loss of confidence in "science," illustrates particularities of the British science-society relationship, as elaborated in Jasanoff, *Designs on Nature.*
22 Department for Business Innovation and Skills, "Principles on Scientific Advice to Government Published for Consultation," December 15, 2009, http://nds.coi.gov.uk/clientmicrosite/content/Detail.aspx?ReleaseID=409612&NewsAreaID=2&ClientID=431.
23 Sheila Jasanoff, "Restoring Reason: Causal Narratives and Political Culture," in Bridget Hutter and Michael Power, eds., *Organizational Encounters with Risk* (Cambridge: Cambridge University Press, 2005), pp. 209–232.
24 Jasanoff, *Designs on Nature,* p. 249.
25 BBC News, *On This Day,* May 16, 1990, http://news.bbc.co.uk/onthisday/hi/dates/stories/may/16/newsid_2913000/2913807.stm.
26 See, for example, "Mad, Bad and Dangerous," *The Economist,* May 25, 1996, p. 15.
27 Lanchester, "New Contagion," 81.
28 "What the Cabinet Will be Eating," March 22, 1996, *Independent,* 5.
29 See, for instance, G. Scarfe, "What Have We Been Fed?" *The Sunday Times,* News Review, March 24, 1996, p. 5.
30 Yaron Ezrahi locates this shift in the democratic gaze far earlier in time, and relates it to the new capabilities of virtual witnessing associated with the scientific revolution. See Ezrahi, *Descent of Icarus: Science and the Transformation of Contemporary Democracy* (Cambridge, MA: Harvard University Press, 1990). Persuasive as his argument is, it is also clear that this shift happened with considerable friction, over extended periods of time, and by different pathways in different political systems. These observations reinforce my own plea for a more historically grounded, more reception-focused, comparative analysis of the democratic politics of knowledge.
31 See, for example, Frank Rich, "The New Rove–Cheney Assault on Reality," *New York Times,* March 13, 2010. Commenting on the political operative Karl Rove's new memoir, which exonerates the Bush administration of all wrongdoing, Rich noted: "Its spin is so uninhibited that even 'Brownie, you're doing a heck of a job!' is repackaged with an alibi."

32 Statement by Professor Sir Richard Southwood for presentation to the European Parliament, Investigative Committee on BSE, Strasbourg, December 9, 1996 (personal communication to the author).
33 On this theme, see Alan Irwin and Brian Wynne, eds., *Misunderstanding Science? The Public Reconstruction of Science and Technology* (Cambridge: Cambridge University Press, 1996).
34 Jasanoff, "Restoring Reason."
35 National Science Foundation, *Science and Engineering Indicators 2006*, ch. 7, "Science and Technology: Public Attitudes and Understanding."
36 The science journalist Daniel Greenberg has written scathingly of US scientists' persistent blaming of a scientifically illiterate public for their imagined woes. Greenberg finds no association between PUS and science funding. See Greenberg, *Science, Money, and Politics: Political Triumph and Ethical Erosion* (Chicago, IL: University of Chicago Press, 2001), pp. 205–233.
37 See in particular Brian Wynne, "Public Understanding of Science," in Sheila Jasanoff, Gerald Markle, James C. Petersen, and Trevor Pinch, eds., *Handbook of Science and Technology Studies* (Thousand Oaks, CA: Sage, 1995), pp. 380–392.

2 The politics of non-knowing

An emerging area of social and political conflict in reflexive modernity

Ulrich Beck and Peter Wehling

1 Introduction: the emerging politics of non-knowing

Before the financial crisis, the political and economic experts pretended to know everything; in the financial crisis, they suddenly know nothing any more (without really admitting this to themselves and to the public). The crisis of the globalised financial markets has again brought home in a dramatic way to a both amazed and deeply distraught public that, especially in the self-proclaimed knowledge societies, phenomena and dynamics of non-knowing are acquiring an importance that is difficult to overestimate as the scale of the threat emanating from civilisation increases. Who – apart from a handful of Cassandras who were mostly dismissed as mavericks and 'prophets of doom' – foresaw, or even had an inkling, that within a short space of time the financial sector would experience dramatic collapses across the globe (Beck 2009), that major banks could be prevented from going under only through state aid on a gigantic scale and that even whole states could be rescued from bankruptcy only with difficulty? In retrospect it turned out that the actors who made such a show of their knowledge in the financial markets did not know what they had got themselves into with the so-called innovative financial products. At any rate, they were incapable of assessing the associated risks. The financial crisis is not the only example which illustrates the explosive power of what is not known in contemporary societies. The threatening, man-made climate change, too, and the potential, but unknown consequences of the release of genetically modified organisms (GMOs), of the spread of 'swine flu' viruses and of the diffusion of environmental chemicals throughout the world emphatically underline that, notwithstanding all assertions to the contrary, numerous spheres of action and politics in contemporary societies are conditioned by non-knowing rather than by knowledge.[1] Especially in a world of delimited threats – world risk society – we are compelled to act under conditions of more or less non-knowing: that is the message which has the significance of a ticking political time bomb.

It is not simply a factual, 'objectively' given lack of relevant knowledge that is secretly propelling the dynamics of ostensible knowledge societies, however, but above all emerging social debates and conflicts concerning the recognition, definition, evaluation and communication of what is, or is supposed to be, not known. Are the limits of knowledge routinely and tacitly neglected or even consciously denied – or are they openly admitted and taken into account? How is ignorance or non-knowledge to be understood and communicated: as a merely temporary information deficit or as a persistent and even irreducible inability to know? Are we faced with known unknowns or with completely unknown and unforeseeable unknowns? And on which of these contrasting framings could and should we rely when we are dealing with the unknown in politics, economy and science?

Such questions are driving a *politicisation of non-knowing* in reflexive modernity whose primary manifestation is the clash between divergent interpretations of what is not known in public conflicts (Beck 1994, 1998; Wehling 2007). Whereas the one side is willing to concede at most manageable knowledge deficits, the other fears that previously unknown and unpredictable environmental and public health harms will occur; for example, through the release of GMOs. While the one side points to apparently clearly known causal connections between the use of fossil fuels and global warming, between smoking and the spread of lung cancer, and calls for appropriate countermeasures, the other side emphasises the enduring uncertainties and gaps in knowledge in order to prevent political interventions and regulation.[2] Clearly non-knowing or 'ignorance claims' (Stocking and Holstein 1993) can be employed deliberately and strategically to achieve specific goals and to promote interests. In other contexts, non-knowing is even being increasingly appealed to as a positive value, above all in the 'right not to know' in predictive genetic testing. This right is supposed to prevent people from being hampered by knowledge concerning future health risks without having access to reliable options for prevention and therapy. Among the questions which arise in this context are: How much do individuals want and need to know about themselves and how much may (or must) interested third parties (employers, insurance companies, state institutions) know about these people and their dispositions to, and risks of, illness? Of course, such debates and controversies do not take place in a power vacuum. On the contrary, it is extremely important who acquires the public power of definition over what is not known, its scope, its relevance and its possible consequences. This holds both in the domestic and international spheres and can refer to political and to subpolitical actors. One must distinguish in this context between strategies for *governing non-knowing* and strategies for those who seek to *govern through non-knowing*, like Foucauldian power strategies, for instance, on global financial markets or in the realm of climate politics. Yet, as far as non-knowing speaks to our inability to anticipate what the future holds, there are *limits to governance*. These limits undermine and disrupt the efforts of governing (through) non-knowing in diverse and interesting ways.

The emerging politics of non-knowledge is, of course, intimately bound up with the politics of knowledge, though it cannot be reduced to the latter. The politicisation of non-knowledge differs from the politics of knowledge primarily in the elusiveness and 'absence' of its object. Non-knowing, by contrast with knowledge, is not a 'tangible' good which can be patented and traded, and over whose possession and use conflicts can arise. It is in the first place 'merely' a definitional construct, a social self-attribution or attribution by others, which asserts the inadequacy and incompleteness of what is known concerning external phenomena and reality. However, because conflicts concerning the nature and scope of non-knowledge refer to the real or presumptive *lack* of knowledge, they cannot be conducted, much less ended, by appealing to available empirically established facts. Thus they quickly veer into the domains of the hypothetical, the seemingly speculative and the paradoxical, and just this is a crucial feature of their political dynamics. By definition, the question of whether thus far unknown and undiscovered negative effects of GMOs 'exist' cannot be answered on the basis of available empirical data, nor can it be decided conclusively by scientific authority. For even if new causal connections were to be discovered, the basic question would remain open whether *yet further* phenomena exist about which we know nothing, and hence which we cannot investigate systematically because we are unaware about where, when and how they will manifest themselves.[3] Such problems are characteristic of the domain of the politics of non-knowing, and it is not surprising that this calls for the creation of new political arenas, new forms of public debate and new decision rules which as yet exist only in rudimentary forms.

In what follows, we would like first to trace the historical process by which a certain interpretation of non-knowing became established in modernity and, against this background, to depict the forms and consequences of the contemporary politicisation of what is not known. In this connection, we will introduce some differentiations of non-knowing which facilitate the analysis of how the perceptions of non-knowing are currently pluralised. Following on this, we would like to illustrate specific forms of the politics of non-knowing and their respective dynamics, using the four examples of global financial crisis, climate change, genetically modified organisms (GMOs) and predictive genetic testing. In a concluding review, we will make a twofold plea for a politics of recognition of non-knowing: on the one hand, non-knowing must be explicitly recognised as an enduring and central condition of action under conditions of reflexive modernity; on the other hand, the variety and plurality of interpretations of what is not known must also be regarded as legitimate and used as a resource of action. In other words, there is not just one 'correct', scientifically certified way of dealing with non-knowing. One consequence which should be drawn from this is the creation of social fora and political arenas which would enhance the democratic accountability of the politics of non-knowing in world risk society.

2 The politicisation of non-knowing – backgrounds, forms and consequences

From the seventeenth century onward, with the emergence of modern science, a new kind of cultural interpretation of non-knowing has become established which has been of major legitimatory and motivational importance for the dynamics of knowledge in modern societies. This new interpretation involves two interconnected aspects. Zygmunt Bauman has characterised the first in very forceful terms as the *temporalisation* of non-knowing (or ignorance):

> Instead of paralysing action, ignorance prompts more effort and boosts the zeal and determination of actors. Ignorance is a not-yet conquered territory; its very presence is a challenge, and the clinching argument of any pep talk summoning support for the next attack in the indeterminable, yet always confident of the ultimate victory, offensive of reason.
>
> (Bauman 1991: 242)

Thus what is not known is always regarded merely as that which is *not yet* known; nothing in the world is held to be unknowable and hence the triumph of knowledge is only a matter of time. In this way ignorance or non-knowing are defined in advance 'as another feather in science's cap. Its resistance is significant solely for the fact that it is about to be broken' (ibid.). This temporalisation of non-knowing can be traced historically to Francis Bacon's grounding of modern science at the beginning of the seventeenth century. An examination of Bacon reveals that this interpretation was not simply a consequence of the liberation of a 'natural' human curiosity from the shackles of religious authority. Rather it was itself shaped by religion – both by the goal of human domination over nature through the accumulation of knowledge as 'a precondition of the recovery of paradise' (Blumenberg 1983: 232) and by the assurance that God concealed the secrets of nature only in order that they could be discovered by human beings.

The second element of the classical modern interpretation of non-knowing consists in its *moral devaluation.* If there is nothing in the world that cannot be known in principle, then non-knowing must be attributed to the inadequacy of the efforts of human beings, whether they attempt to explore nature with questionable methods or they remain mired in settled and comfortable habits of thought. This forms the starting point in Bacon's *Novum Organum* of 1620 for the famous doctrine of the 'idols'; that is, of socially conditioned prejudices and habits of thought which hold the human mind captive and prevent it from gaining access to truth (Bacon 1990: 101ff.). Because and insofar as non-knowing is a result of submission to the idols, it is seen as self-inflicted and a negative moral valuation is attached to it. Not only can it be overcome in principle but it *must* be overcome so that science can contribute to the domination of human beings over nature. The astronomer Johannes Kepler formulated this imperative to overcome non-knowing in particularly drastic terms: 'So long as the mother, Ignorance, lives, it is not safe for Science, the off-spring, to divulge the hidden cause of things' (quoted from Proctor 2008: 30).

The twofold interpretation of non-knowing, as temporalised not-yet-knowing and as self-induced ignorance, achieves a dominant position in modern societies, though this has not prevented repeated attempts to question it from religious, philosophical and even scientific points of view. Only over approximately the past three decades, however, has there been at least a partial change. This has been brought about by the new kinds of risk conflicts and ecological debates on the one hand, and by new perspectives in philosophy, sociology and history of science on the other. This change is reflected in an increasing pluralisation and politicisation of social perceptions and interpretations of what is not known. While it remains dominant, the classical modern interpretation of non-knowing is no longer the only possible and only conceivable one. Thus in certain contexts it is being contested that non-knowing must always be evaluated in negative terms and that it is only a question of time before it is superseded by reliable and complete knowledge. In addition, it is being questioned whether non-knowing is always merely a 'native or originary state' (Proctor 2008: 4) prior to science. For more recent developments in philosophy and sociology of science have shown that non-knowing can also be a *consequence* of scientific knowledge and its technological application. We speak in this sense of *manufactured non-knowing* (Beck 1999, 2009).[4] For science is by no means always capable of anticipating or observing the effects of its interventions in the world. Moreover, since the acquisition of scientific knowledge rests necessarily on selective observations of limited scope, non-knowing is in the process always co-produced along with knowledge (on this see already Fleck 1935), even if this fact usually remains latent and undetected. Thus non-knowing is not merely a result of mental lethargy and bias, of a *lack* of scientific thinking but is simultaneously a product of science itself. Moreover, it is not merely that, with every piece of knowledge acquired, new, as yet unanswered questions also arise, as Karl Popper, for example, repeatedly emphasised. Rather, science, as the example of the 'ozone hole' demonstrates, often does not even know what it does not know and where it should direct its attention and research. Its non-knowing remains latent, implicit and unrecognised – it does not open up any new horizons of enquiry, but often becomes apparent and 'visible' due only to completely unexpected events with potentially grave consequences.

The pluralisation and politicisation of non-knowing represents an extremely complex social and cultural phenomenon. It can be captured more systematically if central importance is accorded to the *relations of definition* of (non-)knowledge as *relations of power* (Beck 2009). 'What the "relation of production" in capitalist society represented for Karl Marx, "relations of definition" represent for risk society.... They form the legal epistemological and cultural power matrix in which risk politics is organized' (Beck 2009: 31–32). This means that the question of which definitions of the unknown are socially accepted and become dominant largely depends on different actors' command of powerful resources such as scientific or political credibility, access to mass media, scientific journals and regulatory institutions, or the ability to do research and produce scientific results. Given this background, we distinguish three dimensions along which social actors contrastingly define and appraise what is not known: the epistemic dimension of the *awareness* (or unawareness) of non-knowing (section 2.1), the social dimension of the *intentionality* of non-knowing (2.2) and the temporal dimension of the *persistence or reducibility* of non-knowing (2.3) (for a more detailed account see Wehling 2006: 116ff.). Occurrences of non-knowing are conceived, defined and interpreted in different ways in each of these dimensions and are thus rendered open to political negotiations, scientific controversies and social struggles.

2.1 The (un-)awareness of non-knowing

Often we have a pretty clear understanding of what we do not know, whereas in other situations we have 'no idea' of what is unknown to us. Hence non-knowing can be differentiated according to the extent to which social actors are aware or unaware of it. Expressed in ideal typical terms, in this epistemic dimension of non-knowing explicitly known, exactly specifiable gaps in knowledge[5] contrast with complete unawareness of what one does not know. Whereas in the former case one can enquire or conduct research in a more or less systematic manner, in the latter it even remains unknown *that* one does not know something and *what* one does not know. In this case all further enquiry is bound to come to nothing as long as one does not have at least a rough idea of what one is looking for. Under such conditions the modernist confidence, which claimed that our ignorance is only temporal and that the unforeseen effects of technical innovations would become known 'in good time' enabling us to intervene and correct them, simply evaporates. Cases such as the 'ozone hole' have made it clear that even the retroactive knowledge and/or causal attribution of damage which has already occurred depends on extremely demanding presuppositions and is often successful only thanks to favourable circumstances.

Against this background, controversies over possibly unknown non-knowing, or *unknown unknowns* (see Kerwin 1993; Beck 1999, 2009; Grove-White 2001; Wynne 2001, 2002; Wehling 2006; Böschen *et al.* 2010) acquire particular explosive power in the social conflicts concerning technological or systemic risks. For in situations of 'negative evidence' (Walton 1996), where knowledge can be acquired only from the *lack* of empirical data, it is ultimately impossible to distinguish positive knowledge from unknown unknowns: Do we know that a new technology (for instance, the genetic modification of organisms) will have no harmful effects if we have (as yet) no empirical evidence of this? Or does this merely mean that we 'have no idea' of where, within what time frames and in what form negative effects could transpire – assuming that they have not already occurred but have not yet been detected?

> The more thorough the search has been, the more we can say that the outcome is no longer just ignorance, but positive knowledge that the thing does not exist. But in many cases, in the middle regions, it could be hard to say whether what we have is ignorance or (positive) knowledge.
>
> (Walton 1996: 140)

It is first and foremost in such 'middle regions' that the monopoly that science has enjoyed until now over interpretation of what is not known is being contested and is beginning to totter. Science can never provide sufficient guarantees to show that the search for unknown effects is complete or that the spatial and temporal horizons of observation chosen in the process are appropriate (see section 3). And if we do not know what we are supposed to be looking for, and even *whether* something unknown exists which calls for and justifies further systematic enquiry, the certainty that our ignorance is always merely temporary and provisional dissolves. At the same time, the notion of a clear and sharply drawn 'boundary' between knowing and non-knowing becomes blurred; it ultimately becomes a *political* question whether we interpret a given situation (the release of GMOs or nano-particles into the environment) as one of knowledge or of non-knowing – and act accordingly.

2.2 *Intentionality of non-knowing*

In the dimension of intentionality, non-knowing is differentiated according to the degree to which it seems to be attributable to the actions or omissions of individuals, social groups or organisations. To put it in ideal typical terms, here the explicit, conscious rejection of certain information ('unwillingness to know') by social actors stands in contrast to completely unintended, and hence apparently 'unavoidable' non-knowing. Intentional non-knowing can be further differentiated into cases in which the ignorance is *one's own* (a result, for instance, of lack of interest or of self-imposed taboos) and those which involve efforts to keep *others* ignorant, for example, through passing on information selectively, through deception techniques, censorship, secrecy, etc. The 'agnotological' questions of the 'diverse causes and conformations' of non-knowing are situated primarily in this sphere of intended non-knowing (Proctor 1995, 2008; see also Stankiewicz 2008).[6] However, it must be kept in mind that, although non-knowing, as 'an actively engineered part of a deliberate plan' of certain actors (Proctor 2008: 9), presumably occurs more often than one would conjecture in modern, 'open' societies, it nevertheless represents just one aspect of the phenomenon. In addition, forms of unseen ignorance play an important role which result from a lack of attention or interest but are not consciously wanted or deliberately produced. Thus the notion of intentional non-knowing is not confined to the *explicit* intention of certain actors to refrain from knowing something, but includes as well those cases in which non-knowing, though not deliberately manufactured, may be causally attributed to the actions and omissions of persons or groups. Yet, it is obviously always contestable whether actors could in fact have known more in certain situations if they had only been more interested in acquiring knowledge.

In this dimension of non-knowing, pluralisation and politicisation comprise two seemingly opposed aspects. On the one hand, different interpretations arise concerning what actors in a given situation could or should have known. Was the 'thalidomide scandal' of the mid-twentieth century 'unavoidable' or could the manufacturer of the sleeping pill have discovered on the basis of comprehensive tests that the supposedly harmless active ingredient causes serious deformities in human foetuses (see Kirk 1999)? On the other hand, it is becoming a matter of controversy, at least in some spheres of social action, how much one should know and what one should better not know. The result is a partial positive revaluation and reappraisal of intentional non-knowing (see Townley 2006) which is diametrically opposed to the modern will to knowledge. The most conspicuous expression of this reappraisal is the by now widely recognised 'right not to know' in predictive genetic testing (see Section 3.4). However, trends towards a revaluation of deliberate non-knowing may be observed even in knowledge management within organisations, one of the core domains of a supposedly knowledge-based economy. Given a scarcely governable abundance of information, conscious ignorance seems to some observers to represent a way of remaining in a position to act and make decisions, with non-knowing even becoming a 'success factor' (Schneider 2006).

2.3 Temporality or persistence of non-knowing

The temporality of non-knowing speaks to the possibility (or impossibility) of transforming non-knowing into knowledge over the course of time. In this dimension, the ideal types of a fundamentally insurmountable 'inability-to-know' or 'inability-ever-to-know' stand over against an always only provisional 'not-yet-known'. As already indicated, the latter represents the dominant perception of what is not known in modern societies and modern science, whereas ideas concerning a non-knowing which is irreducible *in principle* are generally regarded as metaphysical or religious residua. In fact, the modern temporalisation of non-knowing continues to enjoy a high level of persuasiveness in contemporary societies. Nevertheless, a pluralisation and politicisation of interpretations is also discernible with regard to the temporality of non-knowing. It is objected, for example, that the behaviour of ecological and technological systems or of globalised financial systems cannot be predicted or steered in principle on account of their dynamics and complexity (see on this also Rescher 2009: 100ff.). And even where unknowability is not assumed *in principle*, the question becomes important of whether knowledge of unforeseen effects of technological innovations can be acquired 'in good time', i.e. *before* the occurrence of serious, and potentially no longer rectifiable damage (on this see already Collingridge 1980). Beyond the quasi-metaphysical assurance that non-knowing is always only provisional, science cannot provide any convincing guarantee of this in advance for the simple reason that it often does not even know what effects it should be looking for and when these become apparent and detectable. Thus in political debates and social conflicts (over nuclear energy, agri-biotechnology, nanotechnology, etc.), it becomes controversial which interpretation should provide orientation. Whereas critics point to unknown unknowns and to the enduring 'unknowability' of complex causal interconnections, the supporters of technologies assume that the relevant gaps in knowledge are specifiable and can be overcome within manageable time-scales.

In all three dimensions, it has been evident for a number of years that 'classical' modern, hitherto firmly established patterns of interpretation are being challenged by new evaluations of what is not known. In the areas of nuclear power, genetic engineering, nanotechnology, climate change or the globalised neoliberal market economy, society itself has become the site of large-scale experiments which appear to generate results that were in excess of reason's capacity to comprehend or manage. Controversial here are the scope, the significance, the causes and the possible consequences of non-knowing. Different 'ignorance claims' (Stocking and Holstein 1993) are in this way becoming both the object of, as well as instruments and resources in, political disputes. From a sociological perspective, the issue is not so much which of these claims concerning non-knowing is the 'correct' one. The decisive issue for the dynamics of the politics of non-knowing is rather that the hitherto dominant, seemingly self-evident interpretations of non-knowing as merely not-yet-known and as morally dubious ignorance are losing their unconditional validity and opening up a space for pluralisation and politicisation. Closely bound up with this is the fact that science can no longer maintain unchallenged its prerogative on interpretation concerning what is not known. Although the established sciences generally still command significant material, cognitive and symbolic resources for formulating influential definitions of what is not known, its relevance and possible consequences, nevertheless their interpretation is ultimately just one among many. In the following section we would like to explore in greater detail the politicisation of non-knowledge and the attendant social conflicts, using the four examples of global financial crisis, climate change, genetically modified organisms and predictive genetic testing.

3 Politics of non-knowing and their dynamics: four cases

How is non-knowing becoming a topic of political controversies and a political resource, and what different dynamics of a politics of non-knowing may be observed? In an attempt to provide a differentiated answer to these questions we have chosen the four examples mentioned – global financial crisis, climate change, genetically modified organisms and predictive genetic testing – because they enable us to illustrate different facets of the politics and politicisation of non-knowing in each case.[7] In the first two examples, the politics of non-knowing still operates largely, though in very different ways, within a 'modernist' horizon in which the importance of non-knowing is played down. The example of the global financial crisis involves *ignored non-knowing*. Although uncertainty and non-knowing where fully recognised and theorised in parts of economic theory, they were more or less consciously ignored by the economic and political players in order to lend their actions the appearance of rationality and to realise their interests.

In the case of climate change, by contrast, non-knowing is not simply ignored but is transformed into a new kind of 'manufactured certainty' (one which does *not* rest exclusively on scientific knowledge), in order to persuade politics and the public and to prompt them to rapid action. We speak of *manufactured* certainty because here we are no longer dealing with an original self-assurance of scientific knowledge convinced of its truth, but with an actively constructed certainty which is the result of the containment of numerous uncertainties.

By contrast, the examples of genetically modified organisms and of predictive genetic testing exhibit a more far-reaching recognition of both non-knowing and the plurality of perceptions of non-knowing. In conflicts over GMOs, especially the form of *unknown unknowns* plays a central role as a political resource for critics of this technology, whereas in the debates over the opportunities and social risks of predictive genetic testing, a remarkable normative revaluation in the shape of the '*right not to know*' may be observed: deliberate ignorance of one's individual genetic constitution is understood as an interest which merits legal protection and calls for political guarantees.

3.1 Global financial crisis: ignoring the vast realm of non-knowing

It might be surprising that, as early as the 1920s and 1930s, non-knowing was discovered by a scholarly discipline which today no longer wants to have anything to do with it, namely economics. It was Knight (1921) and Keynes who insisted at an early date on a distinction between predictable and non-predictable, or calculable and non-calculable, forms of uncertainty. In a famous article in *The Quarterly Journal of Economics* (February 1937) Keynes (1937: 213–214) writes:

> By 'uncertain knowledge', let me explain, I do not mean merely to distinguish what is known from what is merely probable. The sense in which I am using the term is that in which the price of copper and the rate of interest twenty years hence, or the obsolescence of a new invention are uncertain. About these matters there is no scientific basis on which to form any calculable probability whatever. We simply do not know.

However, Keynes' admonition to open up the field of economic decision-making to the unknown unknowns of future systemic catastrophes hidden in normalised practices of risk-taking was completely neglected in the subsequent development of mainstream economics (including mainstream Keynesian economics). Subsequently, uncertainty was mostly understood as a form of knowledge in which the occurrence of future events can be estimated at least in a subjective way, whereas risk has been as objectively predictable and controllable on the basis of probability calculations. Only since the 1980s has what Keynes termed 'uncertain knowledge' been taken up again in decision theory and at the margins of economic theory under the heading of 'ignorance' (Collingridge 1980; Tietzel 1985; Faber and Proops 1993). Ignorance was then understood as a situation *beyond* both risk and uncertainty in which we are completely unaware of *which* events could even occur and *which* consequences of a decision we have to reckon with (Faber and Proops 1993: 113–114). Thus here the issue is not merely, as in Keynes, that we cannot predict the future development of certain variables (the price of copper, the rate of interest and so on), but in addition that potentially entirely new, unexpected factors will come into play.

In mainstream neoliberal economics, which emerged around the same time, however, these considerations have been largely ignored. Since the 1990s, actors on the global financial markets, relying on the irrelevance of non-knowing, began to introduce a variety of 'innovative' financial products (which are now commonly referred to as 'toxic papers') and in part to outsource them to 'shadow banks', in the mistaken belief that they could calculate and master the inherent risks on the basis of supposedly infallible mathematical models. The possibility of unexpected cumulative effects, such as in fact occurred in the current crisis which began in 2007, was in effect excluded and denied. There are good grounds for assuming that precisely this collective ignoring of non-knowing, combined with the faith in the 'invisible hand' of a nevertheless ultimately opaque market, have profoundly influenced the scope of the crisis. Following the outbreak of the crisis a similar picture emerges: governments and economic actors are faced with the awkward problem of having to make decisions about unimaginable billions of dollars, pounds and euros, on the basis of more or less unadmitted non-knowing. How far the measures taken are successful remains, for the time being, open and difficult to foresee. At the same time, the governments find themselves compelled by the pressure of circumstances to make promises concerning control and security which they may be completely unable to keep in the future, given the incalculability of economic and political dynamics.

The purpose of these brief remarks cannot, of course, be to offer a comprehensive explanation of the global financial crisis, whose causes and scale until now even (and especially) economists do not understand. Rather we wanted to make clear that a certain politics of non-knowing – that of more or less intentionally ignoring what is not known (see Section 2.2) – represents an essential element of the global crisis dynamics. This is not a matter of a kind of 'innocent' and inevitable unawareness of non-knowing, but of an interest-driven suppression and marginalisation of knowledge or at least conjectures concerning the vast realm of non-knowing to which economic theory has in principle had access since at least Keynes. Without doubt, such a strategy of *governing by ignoring* can enjoy success for a long time; nevertheless the associated risk is clear – and is gaining momentum over time. For precisely *because* the manufactured illusion of control and accountability is making the economic and political actors progressively more risk-friendly, the scope and the 'explosive power' of ignored non-knowing are increasing. Thus the attempt to govern by non-knowing is always in danger of undermining its own foundation, as already suggested.

3.2 Climate change: the transformation of non-knowing into manufactured certainty

In 2007 the joint award of the Nobel Peace Prize to the UN Intergovernmental Panel on Climate Change (IPCC) and to Al Gore placed the ecological crisis squarely on the global political agenda. Carefully balancing knowledge claims and uncertainties, the IPCC report asserted that the evidence for man-made climate change is 'unequivocal', and thus almost certain. Apparently the climate scientists and the advocates of an effective reduction of greenhouse gas emissions have in recent years gradually acquired the power of definition over knowledge and non-knowing. Attempts to cite still-existing gaps in knowledge and uncertainties concerning the anthropogenic causation of climate change, along with those concerning its scope and consequences, as reasons (or pretexts) for rejecting a climate protection policy have in the meantime lost their legitimacy.[8] This apparently holds even for the United States where the Bush administration, with the support of arguments from conservative think-tanks, was still pursuing such a policy of obstruction into 2008 (see Oreskes and Conway 2008). Remarkably enough, the 'miracle' (as one is almost tempted to put it) of a global consensus concerning the need to reduce greenhouse gas emissions which has been reached in the meantime (a consensus which does not exclude climate heretics and attacks on the scientific credibility of the IPCC but makes them possible) rests for its part less on the open admission of uncertainty and non-knowing than on what we would like to term 'manufactured certainty'.

How has the presumably indisputable fact of man-made climate change been 'socially constructed'? The IPCC, as a new kind of "boundary organisation" (Guston 2000) between politics and science, not only harmonises the interests and expectations of politics and science, but simultaneously systematises and unifies the state of knowledge of international climate science. It does this not least through a kind of boundary work which is intended to differentiate serious science from dubious speculations emanating from the ranks of the so-called climate sceptics. This represents an attempt to formulate clear statements concerning the future development of the global climate designed to motivate the political and economic players to take rapid and effective action.

Nevertheless, given the enormous gaps in knowledge which indisputably exist concerning the future dynamics of such a complex social-ecological system as the global climate, the result is not naive certainty and unambiguous knowledge. Existing uncertainties are instead being transformed into degrees of reliability and probability of the individual assertions and into ranges of future development, for example, of the degree of warming of the Earth's atmosphere to be expected, or of the rise of the sea level, depending on further emissions of greenhouse gases. Thus, for example, the Fourth Assessment Report of the IPCC of 2007 comes to the conclusion that it is 'very probable' that the greater part of the warming observed since the middle of the twentieth century is being caused by emissions of greenhouse gases by human beings. According to the IPCC's criteria of assessment, 'very probable' corresponds to a probability of 90 to 95 per cent.[9] For the rise in sea level, the report estimates a rise of between 18 and 59 cm until the year 2095, depending on the emissions scenario. However, the uncertainties concerning the rise to be expected exceed this margin, because of the lack of knowledge concerning, for instance, processes taking place in the continental ice sheet, as Stefan Rahmstorf, one of the leading German climate scientists, concedes (Rahmstorf 2007: 193). In fact, a rise of up to one metre or even more by the end of the twenty-first century cannot be excluded. Thus, in this kind of 'politics of non-knowing', uncertainties, gaps in knowledge and limits of predictability are not simply denied and ignored, in the manner of neoliberal economists. Rather non-knowing is inscribed into the knowledge, as it were, resulting in the probabilistic qualification of the respective assertions. But at the same time, the doubt and uncertainties ultimately appear negligible because of the quantification in terms of probabilities, so that de facto certainty is 'produced'. Thus the result (and perhaps also the purpose) is that the findings of climate research are perceived as (virtually) uncontestable facts and as a reliable foundation for political action. In this way a factual *political authority* of science is (re-)produced which is by no means unproblematic and is radically questioned in other domains of social action and social conflicts (nuclear energy, genetic engineering, etc.).

Here it is not our intention to evaluate this strategy normatively, but to analyse it as a particular form of the politics of non-knowing. As such, in spite of its recognition in principle of the limits of knowledge, it continues to be shaped by the classical modern premises under two aspects. On the one hand, given the persisting resistance of economic and political lobby groups, many climate scientists and advocates of an effective global climate politics clearly believe that the willingness and ability to take political action can be generated only through 'manufactured certainty'. This view may be justified under the existing political conditions; nevertheless it does not take the step inherent in the so-called precautionary principle of demanding that political action should also be based on uncertainty and non-knowing. In this way it remains to a large extent captive to the paradigm of *prevention* of more or less precisely known events which is characteristic of industrial society, instead of taking *precautionary* measures against uncertain or even unknown risks (see on this Ewald 2002). On the other hand, non-knowing is transformed into an entity which is fundamentally known, in Merton's sense specifiable and, as regards its relevance, possibly even quantifiable, in the estimations and probability calculations of climate science. As a consequence, the unknown unknowns (for instance, unforeseen events or unexpected threshold effects) we mentioned in section 2.1 are more or less marginalised. Clearly this selective perception and probabilistic 'taming' of what is not known is not without risk. The supposedly exact ranges of the future development, specified to the last degree Celsius and centimetre, could prove to be overestimations (which would immediately lend ammunition to critics of effective reductions in greenhouse gas emissions), but they could also prove to be massive underestimations, as illustrated by the example of the rise in sea level. Thus the development could be far quicker and more dramatic than the scenarios of the IPCC assume. Therefore non-knowing, although seemingly absorbed and neutralised by manufactured certainty, remains an essential element of climate policy. It cannot be excluded that in future it will be thematised by social actors in the radicalised variant of unknown unknowns, as is already the case in the conflict over agrobiotechnology.

3.3 The conflict over GMOs: politicising the unknown unknowns

In hardly any other risk and technology conflict in recent times has non-knowing achieved such prominent and explicit importance as in the controversies over the growing and release of GMOs in agriculture. Moreover, apart from the debates concerning nuclear energy, if at all, hardly any other conflict has witnessed such a far-reaching shift of definitional supremacy over risks and non-knowing in favour of the critics of technology as that which may be observed since the middle of the 1990s, at least in most of the EU member states. Our thesis is that this development is closely bound up with the fact that in the controversy over GMO's there has been an unprecedented politicisation of contrasting interpretations of what is not known, above all of the inability-to-know and, even more so, of the unknown unknowns as opposed to known and specified knowledge gaps (see Grove-White 2001; Wynne 2001, 2002; Böschen *et al.* 2006, 2010). The scientists participating in the development of genetically modified plants and the associated companies (and also initially the political players) stressed over and again towards the public that cultivating these plants is absolutely risk-free and that the few outstanding gaps in knowledge could be rapidly closed through systematic research. Nevertheless (or precisely because of this), critics from among the NGOs, and in part also from science, succeeded in permanently unsettling the long-dominant interpretation of non-knowing as 'specified ignorance' and not-yet-knowns. Against the background of the BSE crisis, which came to a head in Great Britain around the same time as the conflicts over GMOs became radicalised, these players managed to cast doubt upon and discredit the official assurances of security. One of the main ways they achieved this was by introducing the unknown unknowns into the debate; that is, spheres of possible non-knowing beyond the established scientific horizons of perception (see also Magnus 2008). The critics were able to lend plausibility to this interpretation of non-knowing not only by appealing to earlier examples, such as asbestos or CFCs and the ozone hole, apart from BSE itself (see EEA 2001), but by referring to specific aspects of the release of GMOs. Once introduced into the environment, transgenic organisms can de facto no longer be withdrawn, their horizons of operation are almost unlimited in space and in time and the possible directions taken by the effects of genetic manipulations (gene transfer, effects on 'non-target organisms', etc.) are almost impossible to foresee in advance. In view of this, the conviction took root among large sectors of the European public that knowledge concerning the use of GMOs and their effects on the environment and human health is insufficient and that a vast landscape of risk-entailing unknown unknowns extends beyond the domain of what is known to be unknown. An additional factor was that, in view of the complexity and unpredictability of possible harms, the routine expectation 'that if there were harmful effects, evidence would emerge of its own accord and in good time for corrective action' (EEA-Editorial Team 2001: 172) was undermined.

In the conflict over GMOs, institutional reactions, above all in the shape of so-called post-market environmental monitoring, may be observed which, on the one hand, recognise the pluralisation of perceptions of non-knowing and, on the other, offer initial pointers for new ways of dealing with what is not known. The goal of the ten-year monitoring, obligatory in the EU, following the introduction of a transgenic plant is to establish possible threats to the environment or to public health which could *not* be detected in the preceding security and risk research, but which might become apparent only after the release into the environment. In this way the relevance of unknown unknowns for the debate over GMOs is accorded institutional recognition; that is, that the known and calculable risks, as well as the foreseeable uncertainties, possibly comprise only a small proportion of the potential effects and harms. Moreover, the legitimacy and rationality of different, divergent constructions and evaluations of what is not known in science and society is also at least implicitly recognised. However, even post-market environmental monitoring remains captive in one central point to the classical modern temporalisation of non-knowing, namely in its tacit assumption that the potential negative effects of GMOs on human health or the environment can be observed and causally ascribed, and hence also rectified, within the limited, standardised time frame of ten years. Clearly this presupposition is anything but trivial in a situation in which it is not known where, when and in what form such effects would arise in different cases.

Summarising, one can state that the politicisation of non-knowing, and especially of unknown unknowns, by critics of agrobiotechnology has proven to be a highly effective political resource in the conflict over GMOs. The strengths and weaknesses of this resource are two sides of the same coin. By definition, the appeal to possible unknown unknowns cannot be refuted by empirical facts – no more than the existence of unknown unknowns can be empirically justified. However, this should not be interpreted automatically as evidence of the irrationality of this construction of non-knowing. For both epistemological considerations and historical examples illustrate that it is by no means unjustified or irrational to reckon with unknown unknowns beyond the firmly established scientific horizons of observation and expectation. In fact, the at first sight purely speculative appeal to unknown unknowns no longer seems to meet with anything like a complete lack of comprehension and uniform rejection in reflexive modern societies. Instead, in recent years there has been at least a partial recognition of the legitimacy and relevance of different non-knowledge claims, in the EU at any rate, in the wake of risk and technology conflicts.[10] Nevertheless, references to unknown unknowns, or to an inability in principle to understand and control complex causal interconnections, are not simply replacing the hitherto dominant interpretations of non-knowing, but stand in tense and conflictive relations with the latter.

3.4 Predictive genetic testing: the positive revaluation of intentional non-knowing

As already mentioned, the politicisation of non-knowing also places in question the established normative valuations of knowledge and non-knowing. Knowledge was regarded (and of course is still regarded) in modern Western societies as desirable in itself and as having a more elevated moral status than non-knowing. Nevertheless, an astonishing reappraisal and positive revaluation of non-knowing, and in particular of intentional non-knowing, are becoming apparent at least in certain domains of action (see section 2.2). Without doubt the most incisive and interesting example of this is the so-called 'right not to know' in predictive genetic testing (Chadwick *et al.* 1997). Such a right is supposed to prevent individuals from being openly or subtly forced to acquire genetic knowledge concerning possible future diseases. Such knowledge could not only be burdensome, but, in addition, could expose those affected, as 'presymptomatically ill', to the threat of 'genetic discrimination' on the labour market or by insurance companies (see Geller 2002; Geller *et al.* 2002; Lemke 2006).

Predictive genetic testing is geared to detecting dispositions to specific illnesses through the analysis of the individual's genome. The focus here are both seldom-occurring mono-genetic illnesses, like Huntington's disease, which are apparently triggered by a single genetic deviation, and such widespread 'endemic illnesses' as cancer, Alzheimer's disease, diabetes, etc. In the latter cases, genetic factors constitute only *one* element in the complex aetiology of the disease, alongside environmental factors and lifestyle influences. In these cases, DNA tests can establish at best statistically increased risks of becoming ill; it remains unclear, however, whether the person affected will actually become ill, and even if so, when and how serious the illness will be. Moreover, for many of the afflictions mentioned there are either no effective preventive and therapeutic measures (as in the case of Alzheimer's disease) or these are themselves highly invasive, as with prophylactic breast amputation in the case of a genetically increased risk of breast cancer. But also in the case of diseases whose occurrence can be predicted with almost 100 per cent certainty, as in the case of the fatal Huntington's disease, there is often almost a complete lack of preventive or therapeutic possibilities. Thus predictive genetic testing confronts the individuals concerned with a probabilistic biomedical knowledge of future illnesses which, in the most unfavourable cases, offers them grim prognoses, yet without making available possible courses of action in the form of improved prevention or therapy. A significant number of people, especially from so-called risk families, therefore decline to have genetic tests performed. Against this background it is not surprising that, in connection with predictive genetic testing, the at first sight peculiar, if not 'antimodernist' idea of a *right* not to know has gained increasing legal and political acceptance.

An aggravating factor is that knowledge of the genetic dispositions of persons is potentially highly interesting for certain institutional players, for example, life insurance and health insurance companies and employers. Potential customers or employees with statistically elevated risks of illness could be exposed to severe forms of discrimination, whether it be that they would have to pay higher insurance premiums or that they would find only temporally employment or none at all.[11] Thus the right not to know is not a claim restricted to the action of the individuals in question, but calls for legal and political regulations for its protection. Political decisions must be taken, for example, concerning whether and, if so, under what conditions insurance companies or employers can demand a predictive genetic test, or the disclosure of the results of an earlier test before finalising an insurance or employment contract. While this is permitted in some countries, albeit with restrictions, in others it is completely forbidden.

The case of the right not to know in the context of predictive genetic testing provides a highly illuminating example of the emerging politics of non-knowing under two aspects. On the one hand, a normative revaluation has taken place here which no longer discredits the deliberate refusal of possible knowledge ('unwillingness to know') as moral failure but, on the contrary, acknowledges that the option of not knowing, of refusing even scientific knowledge is an important individual resource and a legal interest which merits protection against those who strive to use predictive genetic knowledge in order to meet their economic ends. On the other hand, how far the protection afforded by this right should extend, and to what extent it restricts the interests of other actors, are becoming matters for direct political regulation. In this way, intentional non-knowing, its legitimacy and scope, constitute a new object of political and social conflict. It is to be assumed that the associated political controversies will continue and that their outcomes remain open. In particular, new scientific findings which, for example, would render the DNA tests more informative and reliable, would enhance preventive or therapeutic possibilities or would widen the circle of illnesses which can (in fact or supposedly) be explained by genetic deviations, may again challenge or even undermine the established legal-institutional determinations of the right not to know.

4 Conclusions: towards recognising the politicisation of non-knowing

As the four examples presented above show, one can speak in a twofold sense of a politics of non-knowing. First, non-knowing as well as ascriptions of non-knowing have become important resources in social controversies. One can appeal to absent knowledge to explain and to justify one's own way of acting, or, by contrast, one can ascribe non-knowing to other persons, groups or organisations in order to win opportunities for criticism and to strengthen one's own position. And, of course, through secrecy and censorship one can also pursue one's own interests and goals, just as one can try, through consciously ignoring (as in the case of the right not to know), to protect oneself from burdensome knowledge or simply from "information overload". Second, the politics of non-knowing comprises definitional struggles between social actors concerning which interpretation of what is not known is appropriate in each case, struggles aimed at the acquisition of public power of definition. This definitional dimension is also at least implicitly addressed where non-knowing is employed as a resource of criticism or justification. Someone who legitimises or excuses her conduct by an appeal to non-knowing tacitly assumes that it was a case of unavoidable non-knowing – there was simply no way to know. By contrast, someone who accuses others of non-knowing generally implies that their lack of knowledge was not unavoidable, but rather intentionally generated, be it that the actors concerned more or less consciously ignored their non-knowing or, at least, neglectfully failed to acquire adequate knowledge. With a somewhat different emphasis, it is argued (for instance, by critics of agrobiotechnology or nanotechnology) that with regard to complex interconnections and long-term effects non-knowing is in any case unavoidable, so that the introduction of new technologies whose consequences we cannot anticipate would be irresponsible. In such constellations, the politics of non-knowing is generally closely bound up with the politics of knowledge. The appeal to non-knowing can serve, on the one hand, to reinforce one's own claims to knowledge or to justify corresponding research projects and, on the other, to undermine the (alleged) knowledge of others, to represent it as incomplete and to delegitimise it.

Under these conditions, the modern assumptions concerning non-knowing forfeit their apparently self-evident validity. What is not known is not perceived under all circumstances as merely *not yet* known, nor can the refusal of knowledge be conceived of straightforwardly as morally questionable and as something to be overcome. This does not mean that henceforth, new (presumptive) certainties take the place of the old ones: the modern conviction that everything in the world is knowable is not simply replaced with the equally questionable idea that the world is unknowable in principle. Instead, there is a slightly growing awareness of the fact that knowledge is a contingent and fragile achievement which itself produces non-knowing as its "shadow-side" (Stocking 1998; Wehling 2006). Moreover, non-knowing is by no means in all cases a positive resource of action that deserves protection, but can in fact have disastrous consequences. Thus how we can and should respond to what is not known and to the various constructions of non-knowing has to be negotiated ad hoc, and often under conditions of cognitive and normative disagreement. Here we can appeal to the presumptive authority of science only to a very limited degree. For many examples show that the sciences often do not even know what they do not know, or completely underestimate the relevance and explosive power of what is not known – with the result that many individuals and groups lend scarcely any credibility to the routine safety promises. The politics of non-knowing in reflexive modernity accordingly means recognising that there are no longer any apparently guaranteed certainties which can provide a basis for decisions concerning how to deal with non-knowing – except the single certainty that there are no certainties, indeed that there is not even a clear and reliable dividing line between knowledge and non-knowing. Thus non-knowing is in fact the always accompanying, if frequently invisible and inaccessible, shadow-side of knowledge. Here, the coincidence of (more or less unacknowledged) non-knowing with global risks (such as climate change, the global financial crisis and terrorism) poses dilemmas of a new kind for state action, which is supposed to guarantee the security of citizens: "the hidden critical issue in world risk society is how to feign control over the uncontrollable in politics, law, science, technology, economy and everyday life" (Beck 2002: 41). One could, of course, also say: "how to feign knowing in order to govern the non-knowing (by non-knowing) – staging a smart, specific, side-effects-free, information-driven utopia of governance" (Valverde and Mopas 2004: 239; see also Amoore and de Goede 2008).

What political conclusions, in the widest sense, can be drawn from these findings? The most important consequence, in our view, should be a *reflexive politics of non-knowing* which must include al least three elements:

- First, the explicit recognition of the fact that non-knowing is an inescapable condition framing social and political action, all the more so in reflexive modern societies pervaded by science and technology which are inclined to present themselves as 'knowledge societies'. However, non-knowing is not simply 'a not-yet conquered territory' (Zygmunt Bauman) which stretches out before us and which, although boundless, can nevertheless be conquered step-by-step. What is required is instead to become aware that non-knowledge is always produced together with the acquisition of knowledge – and by no means exclusively in the form of Mertonian specified ignorance or of what is merely 'not-yet-known' for the time being.
- Thus, second, the plurality, legitimacy and equal status in principle of different perceptions and interpretations of non-knowing must also be recognised. It has become apparent that the temporalisation of non-knowing is a contingent historical premise and postulate of early Western modernity which, although it has not, of course, lost its importance entirely, is nevertheless beginning to forfeit its unconditional and exclusive validity. Competing interpretations of what is not known, as unknown unknowns or as inability-to-know, can no longer be discredited and marginalised as 'irrational' or 'hysterical' against this background. Rather they give expression to an independent social rationality in dealing with what is not known. Something similar holds for the firmly established cultural a priori that equates non-knowing, and in

particular the conscious refusal of knowledge, with something morally deficient. In this regard, too, it must be acknowledged that under certain circumstances there can be good reasons for preferring non-knowing to knowledge – even though a generalised preference for non-knowing cannot be derived from this. Which of these respective interpretations one can and should follow is becoming the topic of context- and situation-specific social conflicts and political decisions.

- Given this situation, reflexive modern societies require, third, an extended repertoire of practices and 'techniques' for dealing with what is not known. Among these are, for example, forms of systematic observation of technological innovations and their effects aimed at learning from the unexpected, such as those postulated within the horizon of a 'real experimental' treatment of interventions into society and the environment (Gross and Hoffmann-Riem 2005; Gross and Krohn 2004). However, since even such concepts tacitly rely on the modern assumption that undesirable effects will become visible and rectifiable in good time, they must be supplemented by the development of 'second-order' indicators of non-knowing, such as those proposed, for instance, for chemicals policy (Scheringer 2002; Böschen *et al.* 2010). These should provide indicators for the extent and relevance of unknown unknowns *before* a possibly uncontrollable and unstoppable experiment with new technologies or financial instruments is initiated. The innovative and provocative or, if you will, 'revolutionary' idea implicit in such a politics of non-knowing consists in declining to use certain forms of knowledge and technologies, not because concrete risks can be specified, but 'only' because the scale of the non-knowing entailed seems too great and too risky. In addition, novel forums of discussion and decision-making are needed which are open to different, contrasting perceptions of non-knowing, thus overcoming the dominance of the modernist assumption that non-knowing is always merely temporal.

The politics of non-knowing, whose contours we have attempted to trace by drawing upon a range of examples, will lead – this much can be predicted – only in rare exceptional cases to clear, definitive and consensual solutions. It is instead constituting itself as a new field of social conflicts over the interpretation and relevance of what is not known, and over how we wish to and are able to deal with it.

Notes

1 The discussion on the topics of ignorance, non-knowledge, uncertainty and so on lacks "an agreed-upon nomenclature" as Michael Smithson (2008: 209) has rightly remarked. Contrary to Smithson or Gross (2007), we do not use 'ignorance', but 'non-knowing' (or 'non-knowledge') as the overarching concept, mainly in order to avoid the moral devaluation that might be linked to ignorance, for instance, in the sense of 'ignoring something'.

2 Against this background the historians of science Robert Proctor and Londa Schiebinger have recently developed a research programme, *agnotology*, which is devoted to the analysis of non-knowing and its cultural production (Proctor and Schiebinger 2008; Proctor 2008). The issues dealt with by agnotology are similar to those which we address in our contribution: "We need to think about the conscious, unconscious, and structural production of ignorance, its diverse causes and conformations, whether brought about by neglect, forgetfulness, myopia, extinction, secrecy, or suppression" (Proctor 2008: 3). However, we do not focus exclusively on the causes of non-knowing but also on the various ways in which it is perceived and construed by social actors and on the political controversies which erupt over definitions of what is not known.
3 Such considerations derive their plausibility and forcefulness from historical examples of persisting and unknown non-knowing (see EEA 2001). The most 'prominent' and far-reaching example of this is probably the so-called ozone hole, the destruction of the ozone layer by chlorofluorocarbons (CFCs). When the industrial production and utilisation of these synthetic substances began in the 1930s, people were completely unaware of their grave side-effects in the upper layers of the atmosphere. Only more than forty years later, in the 1970s, did science begin to get to the bottom of the underlying causal interconnections (Farman 2001).
4 Jerry Ravetz, in the early 1990s, coined the concept of 'science-based ignorance' for this; he uses this term to designate "an absence of necessary knowledge concerning systems that exist out there in the natural world, but which exist only because of human activities" (Ravetz 1990: 217).
5 Robert Merton (1987) has termed this 'specified ignorance', thereby making clear that in science demanding efforts are required to differentiate limited and manageable problems out of the wide, amorphous field of what is not known.
6 In addition to military secrecy, the *agnotology* developed by Proctor deals above all with the tobacco industry's long-standing conscious production of doubt and ignorance around the consequences of smoking (Proctor 1995, 2008: 11ff.). Proctor regards this as a prime example of actively 'manufactured ignorance' (Proctor 2008: 11).
7 These examples speak mainly to the social processing of self-produced economic, ecological and technological risks. We do not mean to imply by this that these are the only contexts in which the politics of non-knowing are emerging. Other important examples are, for instance, the mobilisation of self-attributed non-knowing in dealing with the Holocaust (see Longerich 2006) or the connection between racism and non-knowing (Sullivan and Tuana 2007; Mills 2008).
8 Yet, as the poor results of the Copenhagen Conference in December 2009 demonstrate, this does not mean that the majority of national governments are in fact willing to launch adequate and effective political strategies in order to mitigate global climate change.
9 The third Assessment Report of 2001, by contrast, assessed this as merely 'probable'; this corresponds to an assumed probability of between 66 and 90 per cent.
10 Important in this context may also have been the fact that in recent years it has transpired that, even within science, there exist different but equally well-founded 'cultures of non-knowing'; that is, different forms and practices of interpreting and dealing with what is not known (see Böschen *et al.* 2006, 2010).
11 However, forms of genetic discrimination could also develop within informal social contexts, such as the marriage market. Now that relatively cheap individual genetic tests are increasingly available over the Internet, it is by no means improbable that in the future people will also take account of the genetic profile of potential marriage partners.

References

Amoore, L. and de Goede, M. (eds) (2008): *Risk and the War on Terror.* Abingdon, Oxon/New York: Routledge.

Bacon, F. (1990): *Neues Organon. Lateinisch – deutsch. Hrsg. und mit einer Einleitung von Wolfgang Krohn. 2 Teilbände.* Hamburg: Meiner (Orig. 1620).

Bauman, Z. (1991): *Modernity and Ambivalence.* Ithaca, NY: Cornell University Press.

Beck, U. (1994): The Reinvention of Politics: Towards a Theory of Reflexive Modernization. In: U. Beck, A. Giddens and S. Lash *Reflexive Modernization: Politics, Tradition and Aesthetics in the Modern Social Order.* Cambridge: Polity Press, pp. 1–55.

—— (1998): Misunderstanding Reflexivity: The Controversy on Reflexive Modernization. In: U. Beck *Democracy without Enemies.* Cambridge/Malden, MA: Polity Press. pp. 84–102.

—— (1999): *World Risk Society.* Cambridge/Malden, MA: Polity Press.

—— (2002): The Terrorist Threat: World Risk Society Revisited. *Theory, Culture and Society* 19(4): 39–55.

—— (2009): World at Risk. Cambridge, UK/Malden, MA: Polity Press.

Blumenberg, H. (1983): *The Legitimacy of the Modern Age* (translated by Robert M. Wallace). Cambridge, MA: MIT Press.

Böschen, S., Kastenhofer, K., Marschall, L., Rust, I., Soentgen, J. and Wehling, P. (2006): Scientific Cultures of Non-Knowledge in the Controversy over Genetically Modified Organisms (GMO): The Cases of Molecular Biology and Ecology. *GAIA* 15(4): 294–301.

Böschen, S., Kastenhofer, K., Rust, I., Soentgen, J. and Wehling, P. (2010): "Scientific Non-Knowledge and Its Political Dynamics: The Cases of Agribiotechnology and Mobile Phoning". *Science, Technology and Human Values* 35(6) 783–811.

Chadwick, R., Levitt, M. and Shickle, D. (eds) (1997): *The Right to Know and the Right not to Know.* Aldershot: Avebury.

Collingridge, D. (1980): *The Social Control of Technology.* New York: St Martin's Press.

EEA (European Environment Agency) (ed.) (2001): *Late Lessons from Early Warnings: The Precautionary Principle 1896–2000.* (Environmental Issue Report No. 22). Copenhagen.

EEA-Editorial Team (2001): Twelve Late Lessons. In: EEA (ed.) *Late Lessons from Early Warnings: The Precautionary Principle 1896–2000* (Environmental Issue Report No 22). Copenhagen, pp. 168–191.

Ewald, F. (2002): The Return of Descartes' Demon: An Outline of a Philosophy of Precaution. In: T. Baker and J. Simon (eds) *Embracing Risk: The Changing Culture of Insurance and Responsibility.* Chicago, IL: University of Chicago Press, pp. 273–301.

Faber, M. and Proops, J. (1993): *Evolution, Time, Production and the Environment,* 2nd rev. and enl. edition. Berlin/Heidelberg/New York: Springer Verlag.

Farman, J. (2001): Halocarbons, the Ozone Layer and the Precautionary Principle. In: EEA (ed.) *Late Lessons from Early Warnings: The Precautionary Principle 1896–2000.* (Environmental Issue Report No. 22). Copenhagen, pp. 76–83.

Fleck, L. (1935/1979): *The Genesis and Development of a Scientific Fact* (edited by Thaddeus J. Trenn and Robert K. Merton, foreword by Thomas Kuhn). Chicago, IL: University of Chicago Press.

Geller, L.N. (2002): Current Developments in Genetic Discrimination. In: J.S. Alper, C. Ard, A. Asch, J. Beckwith, P. Conrad and L.N. Geller (eds) *The Double-Edged Helix: Social Implications of Genetics in a Diverse Society.* Baltimore/London: The Johns Hopkins University Press, pp. 267–285.

Geller, L.N., Alper, J.S., Billings, P.R., Barash, C.I., Beckwith, J. and Natowicz, M.R. (2002): Individual, Family, and Societal Dimensions of Genetic Discrimination: A Case Study Analysis. In: J.S. Alper, C. Ard, A. Asch, J. Beckwith, P. Conrad and L.N. Geller (eds) *The Double-Edged Helix: Social Implications of Genetics in a Diverse Society*. Baltimore/London: The Johns Hopkins University Press, pp. 247–266.

Gross, M. (2007): The Unknown in Process: Dynamic Connections of Ignorance, Non-Knowledge and Related Concepts. *Current Sociology* 55(5): 742–759.

Gross, M. and Hoffmann-Riem, H. (2005): Ecological Restoration as a Real-World Experiment: Designing Robust Implementation Strategies in an Urban Environment. *Public Understanding of Science* 14(3): 269–284.

Gross, M. and Krohn, W. (2004): Science in a Real World Context: Constructing Knowledge through Recursive Learning. *Philosophy Today* 48(5): 38–50.

Grove-White, R. (2001): New Wine, Old Bottles: Personal Reflections on the New Biotechnology Commissions. *Political Quarterly* 72(4): 466–472.

Guston, D.H. (2000): *Between Politics and Science: Assuring the Integrity and Productivity of Research*. Cambridge: Cambridge University Press.

Kerwin, A. (1993): None Too Solid: Medical Ignorance. *Knowledge: Creation, Diffusion, Utilization* 15(2): 166–185.

Keynes, J.M. (1937): The General Theory of Employment. *Quarterly Journal of Economics* 51(2): 209–223.

Kirk, B. (1999): *Der Contergan-Fall: eine unvermeidbare Arzneimittelkatastrophe?* Stuttgart: Wissenschaftliche Verlagsgesellschaft.

Knight, F.H. (1921): *Risk, Uncertainty and Profit*. Boston/New York: Houghton Mifflin.

Lemke, T. (2006): *Die Polizei der Gene. Formen und Felder genetischer Diskriminierung*. Frankfurt am Main/New York: Campus Verlag.

Longerich, P. (2006): *"Davon haben wir nichts gewusst!" Die Deutschen und die Judenverfolgung 1933–1945*. München: Siedler.

Magnus, D. (2008): Risk Management versus the Precautionary Principle: Agnotology as a Strategy in the Debate over Genetically Engineered Organisms. In: R.N. Proctor and L. Schiebinger (eds) *Agnotology: The Making and Unmaking of Ignorance*. Stanford, CA: Stanford University Press. pp. 250–265.

Merton, R.K. (1987): Three Fragments from a Sociologist's Notebook: Establishing the Phenomenon, Specified Ignorance, and Strategic Research Materials. *Annual Review of Sociology* 13: 1–28.

Mills, C.W. (2008): White Ignorance. In: R.N. Proctor and L. Schiebinger (eds) *Agnotology: The Making and Unmaking of Ignorance*. Stanford, CA: Stanford University Press, pp. 230–249.

Oreskes, N. and Conway, E.M. (2008): Challenging Knowledge: How Climate Science Became a Victim of the Cold War. In: R.N. Proctor and L. Schiebinger (eds) *Agnotology: The Making and Unmaking of Ignorance*. Stanford, CA: Stanford University Press, pp. 55–89.

Proctor, R.N. (1995): *Cancer Wars: How Politics Shapes What We Know and Don't Know about Cancer*. New York: Basic Books.

—— (2008): Agnotology: A Missing Term to Describe the Cultural Production of Ignorance (and Its Study). In: R.N. Proctor and L. Schiebinger (eds) *Agnotology: The Making and Unmaking of Ignorance*. Stanford, CA: Stanford University Press, pp. 1–33.

Proctor, R.N. and Schiebinger, L. (eds) (2008): *Agnotology: The Making and Unmaking of Ignorance.* Stanford, CA: Stanford University Press.

Rahmstorf, S. (2007): Der Anstieg des Meeresspiegels. In: M. Müller, U. Fuentes and H. Kohl (eds) *Der UN-Weltklimareport.* Köln: Kiepenheuer & Witsch, pp. 190–194.

Ravetz, J. (1990): *The Merger of Knowledge with Power: Essays in Critical Science.* London/New York: Mansell.

Rescher, N. (2009): *Ignorance: On the Wider Implications of Deficient Knowledge.* Pittsburgh, PA: University of Pittsburgh Press.

Scheringer, M. (2002): *Persistence and Spatial Range of Environmental Chemicals.* Weinheim: Wiley-VCH.

Schneider, U. (2006): *Das Management der Ignoranz. Nichtwissen als Erfolgsfaktor.* Wiesbaden: Deutscher Universitäts-Verlag.

Smithson, M.J. (2008): Social Theories of Ignorance. In: R.N. Proctor and L. Schiebinger (eds) *Agnotology: The Making and Unmaking of Ignorance.* Stanford, CA: Stanford University Press, pp. 209–229.

Stankiewicz, P. (2008): Invisible Risk. The Social Construction of Security. *Polish Sociological Review* 161(1): 55–72.

Stocking, S.H. (1998): On Drawing Attention to Ignorance. *Science Communication* 20(1): 165–178.

Stocking, S.H. and Holstein, L. (1993): Constructing and Reconstructing Scientific Ignorance: Ignorance Claims in Science and Journalism. *Knowledge: Creation, Diffusion, Utilization* 15(2): 186–210.

Sullivan, S. and Tuana, N. (eds) (2007): *Race and Epistemologies of Ignorance.* Albany, NY: SUNY Press.

Tietzel, M. (1985): *Wirtschaftstheorie und Unwissen. Überlegungen zur Wirtschaftstheorie jenseits von Risiko und Unsicherheit.* Tübingen: Mohr.

Townley, C. (2006): Toward a Revaluation of Ignorance. *Hypatia* 21(3): 37–55.

Valverde, M. and Mopas, M.S. (2004): Insecurity and the Dream of Targeted Governance. In: W. Larner and W. Walters (eds) *Global Governmentality: Governing International Spaces.* London/New York: Routledge. pp. 233–250.

Walton, D. (1996): *Arguments from Ignorance.* University Park, PA: Pennsylvania State University Press.

Wehling, P. (2006): *Im Schatten des Wissens? Perspektiven der Soziologie des Nichtwissens.* Konstanz: UVK.

Wehling, P. (2007): Die Politisierung des Nichtwissens: Vorbote einer reflexiven Wissensgesellschaft? In: S. Ammon, C. Heineke, K. Selbmann and A. Hintz (eds) *Wissen in Bewegung. Vielfalt und Hegemonie in der Wissensgesellschaft.* Weilerswist: Velbrück Wissenschaft, pp. 221–240.

Wynne, B. (2001): Expert Discourses of Risk and Ethics on Genetically Manipulated Organisms: The Weaving of Public Alienation. *Politeia* 17(62): 51–76.

—— (2002): Risk and Environment as Legitimatory Discourses of Technology: Reflexivity Inside Out? *Current Sociology* 50(3): 459–477.

3 Technology, legal knowledge and citizenship

On the care of Locked-in Syndrome patients

Fernando Domínguez Rubio and Javier Lezaun

1 Introduction

On the morning of 12 July 1999, Jose C., then thirty-three years old, suddenly fainted when he was about to take a shower. After days of close monitoring in the hospital, during which he showed no sign of consciousness, the doctors concluded that he had suffered a stroke that had left him in a persistent vegetative state. The medical team informed Jose's wife and family that, given the extensive physiological and neurological damage the stroke had caused, it was highly unlikely he would survive longer than two months.

Over the weeks that followed this devastating diagnosis, Jose's wife, Maria, identified a barely noticeable pattern in the movement of his right-hand index finger. She alerted the medical team, but the doctors dismissed the idea that the movement could be a sign of conscious brain activity, attributing it instead to the sort of spasmodic muscle contraction typical of patients in vegetative states. Four months after the initial stroke, however, and as the result of Maria's dogged insistence, the medical team acceded to perform additional tests to discard the possibility of conscious action. To their surprise, the results showed that despite the damage the stroke had caused to Jose's nervous system he remained fully aware and conscious, capable of hearing, understanding, reasoning and commanding the movement of his right-hand index finger. The medical team reversed its initial assessment and diagnosed him with Locked-in Syndrome.

Locked-in Syndrome (hereafter LIS), also known as coma vigilante, is a rare neurological disorder normally caused by infarct, haemorrhage, or trauma leading to a brainstem lesion. It entails complete paralysis of nearly all the voluntary muscles of the body, except for vertical eye movement. Outwardly LIS patients resemble those in vegetative states, but there is a crucial difference: LIS patients remain fully aware and conscious, with their intellective capabilities intact. Hence the name of the disorder: the individual is locked inside his body, unable to express his consciousness and translate his thoughts and intentions into words or actions.[1]

In 2000, a local court in Spain declared Jose 'totally incapable'. In Spanish law, the classification of an individual as totally incapacitated triggers the 'replacement of his will' by the appointment of a legal guardian that operates under the regime of *tutela* (tutelage) (Código Civil [2008]: Title X). Having found the person unable to govern himself, the court, through a power of attorney, proceeds to transfer his legal *persona* to another person, who, with judicial supervision, then becomes his legal representative. The individual is stripped of some of his civil rights – such as the ability to enter into a contract or administer his property – as well as of most of his political rights, notably the right to vote. The declaration of total incapacity effectively removes that individual from the political life of the community.

In 2004, Jose C. appealed this ruling. Despite his severe physical disability, he argued, his mental faculties remained intact. In a landmark decision, the regional Appeals Court ruled in his favour and restored his voting rights, as well as the rest of the lost legal powers. Jose left the state of tutelage and regained his status as a full-fledged member of the community. A particularly noteworthy aspect of this ruling, and an issue we will discuss in detail below, is that the Court's decision was not motivated by a sudden improvement in Jose's physical condition – which, as is customarily the case in LIS patients, had remained essentially unaltered. The new legal judgment was grounded, rather, in the fact that over the four years that had passed since the initial decision, Jose had gained access to a series of augmentative communication devices, particularly voice reproduction software and a specially adapted computer interface that allowed him to use the Internet. With the help of these adaptive technologies, Jose had become able to translate internal mental states into speech and, to a lesser degree, actions. As a result, he again became intelligible to the legal system as an actor 'capable of governing himself', and thus entitled to the recognition of his full legal *persona*.

In this chapter, we explore the interaction between adaptive technologies, systems of care, and the intelligibility of LIS patients as full legal persons, subjects endowed with the standard range of civil and political rights. Dependent for the display of their self-governing ability on a heterogeneous assemblage of communicative devices, caregivers and, increasingly, computer interfaces, LIS patients and the socio-technical configurations in which they are immersed unsettle some of the biological presuppositions at the heart of Western legal and political systems (Pottage and Mundy 2004; Rose 2007; Strathern 1992). The canonical separation of persons and things, as well as the traditional definition of the human body as the natural container of the person, become problematic assumptions when, as in the case of LIS patients, the production of intelligible *personae* is dependent on the imbrication of bodies, technological devices and extended systems of care. We will focus on the heuristics that characterize legal knowledge as it attempts to apprehend and classify these new mixtures of biological, technical and social components. This will allow us to understand how key political categories, like that of citizenship, are being redefined by the intersection of legal techniques of personification and new assistive technologies.

Our argument will draw on a comparison between the case of Jose C. and that of Mauricio, another Spanish LIS patient who also appealed a ruling of total incapacity after regaining the ability to communicate with others. Unlike Jose, however, Mauricio saw his legal status confirmed by the Spanish Supreme Court, and never recovered his full legal *persona*. Both Jose and Mauricio engaged in arduous attempts to display before the courts the qualities – awareness, communication, self-governance – that the law demands for the attribution of legal personhood. The disparity of their fates draws our attention to the legal mechanisms and practices of care whereby the disabled and mute human body is made to coincide with, or, alternatively, is disentangled from that 'still imprecise, delicate and fragile' legal construct, the category of 'person' (Mauss 1985 [1938]: 1).

The contrasting legal fates of Jose C. and Mauricio will allow us to advance three claims. First, that novel techno-scientific knowledges and devices – whether in biotechnology, reproductive technologies, human–computer interaction or neurocognitive enhancement – blur the biological demarcations traditionally employed by Western legal systems to define the boundaries of the citizen. These knowledges enable the delegation and materialization of internal processes and capacities, such as speech, intentionality or agency, onto different devices and techno-scientific systems, thereby giving rise to distributed forms of personhood.

Second, we suggest that, confronted with bodies thoroughly dependent on distributed networks of people, objects and devices, the law becomes once again the fundamental system of personification. The notion of the citizen as an autonomous, self-contained, self-governing body is still the cornerstone of the legal machinery for the production of persons, but this notion must be deployed flexibly to contend with novel socio-bio-technological hybrids. It is up to the law to make anew the distinction between the socio-material configurations that give expression to forms of being entitled to civil and political rights, and those which fail to produce the kind of projection of personhood the law requires in order to recognize individuals as full-fledged legal subjects. As the discussion of our two cases will show, this process of legal demarcation depends on the kind of evidence that different socio-technical configurations are able to produce: the devices and technologies through which the person is presented can be seen by the courts as vehicles for the expression of a self-governing self, or as screens hiding an ineffable being. The former vision leads to the recognition of the LIS patient as a full legal *persona*; the latter triggers a moment of radical suspicion, when the law refuses to presume an autonomous self behind – or at the centre of – the assemblage of people, devices and technologies through which the patient is re-presented to the world.

Finally, we will explore the implications of these examples of legal (un) intelligibility to a reconsideration of the key juridico-political category at stake, that of citizenship. The jurisprudence on LIS patients offers a tantalizing opportunity to rethink citizenship as a fragile *position* embedded in socio-technical systems and compatible with intense forms of care, rather than as an abstract *condition* grounded in an isolated, self-governing body.

2 Silence, unknowability and communicative aids

LIS has a rather particular history. The first descriptions of the syndrome are found in literature, rather than in medicine. The most famous example is Alexander Dumas's 1844 novel *Le Comte de Monte-Cristo,* where the writer introduces the character of Monsieur Noirtier de Villefort, an old Bonapartist who, after suffering an apoplectic stroke, was left as 'a corpse with living eyes' (Dumas [1844] 1996: 564). Two decades later, in 1867, Émile Zola provided a very similar description of the syndrome in his novel *Thérèse Raquin,* the story of an old widow who suffered a crisis that left her 'a walled-up brain, still alive, but buried in a lifeless frame' (2008: 146). In the term 'Locked-in Syndrome', as in some of its French or Spanish equivalents – 'maladie de l'emmuré vivant', 'síndrome de cautiverio' – one still hears the echoes of this literary genealogy. The condition was first medically identified by a group of Oxford neurosurgeons in 1941 (Cairns *et al.* 1941), but remained without a medical name for two further decades, until the American neurologists Fred Plum and Jerome Posner coined the term 'Locked-in-Syndrome' in the 1966 edition of *The Diagnosis of Stupor and Coma.*

Plum and Posner described there the case of the 'de-efferented' individual, a person 'with no means of producing speech or movement', and defined his condition as a combination of quadriplegia, lower cranial nerve paralysis and mutism with preservation of consciousness, vertical gaze and upper eyelid movement (see also Smith and Delargy 2005). The reason for the late medical definition lies in the difficulty of diagnosis and the outward similitude between LIS patients and those in persistent vegetative states. Given the nearly total paralysis of the patient, there are hardly any behavioural cues from which doctors can infer the presence of a living conscience in the inert body. An individual suffering from classical LIS can only resort to vertical eye movement to convey alertness to others. This signal, however, can be easily misinterpreted as one of the involuntary muscular movements common in patients in comatose states. Furthermore, most LIS patients suffer on the onset of the syndrome forms of neurological impairment that can make eye movement inconsistent or easily exhausted. 'A high level of clinical suspicion' write Plum and Posner, 'is required on the part of the examiner to distinguish a locked-in patient from one who is comatose' (4th edition, 2007: 7). It is indeed often a relative, not a doctor, who first identifies signs of awareness in the patient: a review of forty-four cases of LIS found that in just over half of the cases it was a family member who first realized that the patient was conscious (Leon-Carrion *et al.* 2002). The verification of the diagnosis depends then on the willingness of doctors to perform additional tests to measure brain activity, a task that is often further complicated by other effects of the brainstem injury, such as memory loss, deafness or severely distorted hearing. The cognitive and neurobehavioural assessment of the patient will thus typically evolve over weeks and months, as slight improvements in his state (such as the partial recovery of head movement) allows the medical team to slowly confirm his responsiveness (Smart *et al.* 2008).[2] Even today, when the condition is widely known in the scientific literature and technologies to identify cerebral metabolism are commonly available, LIS is detected on average seventy-eight days after the onset of the condition – and it is not difficult to find cases in which the patient remains misdiagnosed for several years (Leon-Carrion *et al.* 2002).

Right from the start, then, LIS patients are dependent on different technologies and expert knowledge systems for the proper identification of their predicament, and its differentiation from other conditions, such as vegetative or minimally conscious states. Once the patient is stabilized, the paramount task is to facilitate his capacity to translate mental states into words and actions. Different communicative aids, joining the body to diverse constellations of people, artefacts and technologies, are routinely used to enable LIS patients to express their desires and thoughts. The most basic and widely used interface is an alphabet board with the letters of the alphabet written in different rows and columns (see Figure 3.1)

While the board is held in front of the patient, someone points to a letter, noting it down when the patient selects it (for instance, by producing two rapid upward eye movements). This operation is repeated again, letter by letter, until a full word emerges. This laborious form of communication often offers the patient his first opportunity to break out of the imprisoning silence brought on by the syndrome. It is, however, an excruciatingly slow method: the composition of a simple sentence can take several minutes. The process is extremely tiresome to the patient and prone to misunderstandings, repetitions and errors.

<table>
<tr><td>A</td><td>B</td><td>C</td><td>D</td><td colspan="2">End of word</td></tr>
<tr><td>E</td><td>F</td><td>G</td><td>H</td><td colspan="2">End of sentence</td></tr>
<tr><td>I</td><td>J</td><td>K</td><td>L</td><td>M</td><td>N</td></tr>
<tr><td>O</td><td>P</td><td>Q</td><td>R</td><td>S</td><td>T</td></tr>
<tr><td>U</td><td>V</td><td>W</td><td>X</td><td>Y</td><td>Z</td></tr>
</table>

Figure 3.1 Alphabet Board (source: Smith and Delargy (2005)).

Over the years, different patients and medical teams have essayed a variety of alternatives to improve the efficiency of the alphabet board. Jean-Dominique Bauby, a French LIS patient, rearranged the letters of the board by placing the most commonly used first, and was able thanks to this abbreviated method to increase the speed of communication and write his famous book *The Diving Bell and the Butterfly* (Bauby 1998). Other improvements to this communicative interface have included the separation of vowels and consonants in different columns, or the addition of common daily actions and commands to the board, such as 'open', 'close', 'eat', 'give me', etc. Although the introduction of these variations can certainly improve the speed of communication, this type of interface only allows the patient's communication to emerge at a pace and in a form that barely affords the expression of very simple commands and orders.

The limitations and legal implications of this communicative interface were made evident during Mauricio's appeal against the declaration of total incapacity. As we noted earlier, in Spanish law the declaration of total incapability is applied to individuals who, due to inborn or acquired physical impairment or mental illnesses, are unable to govern or care for themselves. The relevant article of the Civil Code is commonly used to characterize individuals in vegetative states, or suffering from neurological or psychiatric conditions such as dementia, schizophrenia or Alzheimer's. Its extension to LIS patients is contentious. Although the syndrome disables their bodies and breaks the connection between internal states and actions or words, patients suffering from LIS preserve their mental faculties unscathed. If they retain their capacity to think and will, but lack the power to actualize these capacities, do LIS patients warrant a declaration of total incapacity?

In 1997 a local court in Betanzos, a city in the northwestern region of Galicia, had reached that conclusion in the case of Mauricio. He immediately appealed the decision, on the basis that, despite his extreme physical disability and after living with LIS for several years, he was still in full possession and control of his mental capacities. He retained the ability to understand speech and text, and to reason autonomously; what he lacked was solely the physical capacity to autonomously display those capacities in the form of words or actions. By declaring him totally incapacitated, Mauricio alleged in his appeal, the court had 'silenced his undoubted mental, cognitive and volitional capacity, as well as his ability to communicate and express his will' (Tribunal Supremo Sala de lo Civil [2004], 584/2000). The claim was backed by medical analyses attesting to the fact that the syndrome had left Mauricio's higher mental capacities intact, and therefore did not compromise his capacity to govern himself. The forensic report considered proven that 'Mr Mauricio is owner of his acts and has sufficient capacity to take decisions in all areas of his life, although he needs the aid of another person for their physical execution' (ibid.). When the Appeals Court, in spite of the medical evidence, turned down Mauricio's appeal, his lawyer took the case to the Spanish Supreme Court, which in 2004 finally rejected his appeal and upheld the initial declaration of total incapacity. Mauricio was to remain under the guardianship of his mother.

One aspect of the Supreme Court's ruling is particularly striking. During the appeals trial Mauricio had tried to demonstrate his ability to communicate by means of an alphabet board. The Appeals Court's decision had included a detailed description of this attempt, a description that five years later the Supreme Court reproduced verbatim in the justification of its decision:

> Showing him a laminated card, the nurse asks D. Mauricio whether he wants to say something, and she slowly points with a pen to each letter of the alphabet; D. Mauricio chooses the desired letter with an affirmative nod, slightly moving his head and eyes downward; next the nurse writes down the chosen letter, and continues pointing to the letters until D. Mauricio chooses the next one, she writes it down again, until D. Mauricio has completed what he wants to say: 'today is Thursday.'
>
> (ibid.)

This description of Mauricio's communicative efforts provides a clue to the reason why the Supreme Court, despite the evidence presented in support of the claim that Mauricio's mental capacities were intact, decided to uphold the declaration of total incapacity. For the Supreme Court the decision on whether Mauricio ought to be classified as 'totally incapacitated' hinged solely on the evidence of his autonomy; on proving, in the Court's (and the law's) oft-repeated phrase, his 'ability to govern himself'. It is not enough, the Court reasoned, to find signs of consciousness, proof of internal volitional or cognitive capacities. The focus ought to be, the Court argued, on the 'factual reality of the person' (ibid.); that is, on whether the person is legible to the law as a self-governing subject in and through the actualization of those internal capacities. In other words, whether Mauricio retained his mental abilities, even his ability to communicate, was a secondary consideration, since the key issue was his ability to evince the 'natural competence to rule himself and administer his property' (ibid.).

Rather than featuring as proof of Mauricio's ability to express an autonomous and intact mental capacity, the laborious production of 'today is Thursday' operated in the Court's ruling as evidence of his incapacity for self-government. His attempt at communication exposed, in the Court's eyes, the degree of mediation involved in eliciting his self, Mauricio's absolute reliance on the actions and interpretations of others to express thoughts, intentions and desires. The board, the card and the pen remained powerless objects unless activated by the nurse, and the Court interpreted this reliance on the actions of 'third persons' as evidence of a lack of autonomy. Mauricio's communication, the Court declared, 'is not spontaneous, is not produced on his behalf, but on behalf of third persons, he thus lacks the liberty to carry out the decisions he has previously adopted, in relation to his own person as well as the administration of his property' (ibid.). This dependency was further aggravated, in the opinion of the Court, by the fact that the interface was only effective for people with previous knowledge of Mauricio's state – that is, when used by people aware that his winks were part of an effort to communicate, and not mere unintentional (and thus meaningless) blinks. And even if the interlocutor were fully cognizant of Mauricio's conscious state, the communicative interface was prone to misunderstandings, repetitions and errors – a wink could easily be mistaken for a blink, and vice versa. The limited portability of the interface is one of the key arguments the Court adduced to justify its decision: this system, it argued, 'might be sufficient for his relationships with the people in charge of his care, but is not enough for a normal communication with his external environment that would allow him to rule himself and his property' (ibid.).

The Court exerted to the full the suspicion that this system of communication represented, or could eventually become, a form of ventriloquism, in which Mauricio's self only emerged through the interpretations of spokespersons. In these circumstances, it was impossible to verify if and when the messages that emerged through the socio-technical system of assistance were an accurate reflection of his actual thoughts and desires, the projection of a veiled but intact and authentic self, or a fraudulent impersonation of Mauricio's self. The social, technical and material aspects of the system in which Mauricio was inserted and through which he spoke – nurse, laminated card, pen – only offered a highly mediated expression of his internal state; they were in fact further proof of a loss of his powers of 'self-government'. The Court brought this point home by repeatedly mentioning in its decision Mauricio's aphasia, and by emphasizing several of the activities of his daily life for which he was fully dependent on intensive systems of care. Permanently reliant on the help and supervision of others to carry out routine physical tasks, it was 'impossible' for Mauricio, the Court declared, to perform 'normal behaviour' (ibid.).

Mauricio's case reveals some of the conditions under which a subject becomes (un)intelligible as a full legal person. By disabling his body, LIS had removed the element that sustained the architecture of relations between self, thought and actions that rendered his self knowable as a self-governing individual. Medical and physiotherapeutic interventions had attempted to reinstate the conditions of intelligibility by complementing Mauricio's impaired body with a new socio-technical system of support and care. Even if the combination of body, alphabet board and nurse enabled Mauricio to communicate his thoughts and be intelligible to others as a living mind, in the eyes of the Supreme Court it still did not allow him to demonstrate the specific form of personhood legal knowledge requires to apprehend an individual as a full legal subject.

We are here reminded of Mauss's famous discussion of the oft-found etymology of the Latin *persona*: *per/sonare*, the mask through which the voice of the actor sounds (Mauss 1985 [1938]). Whether the mask is deemed to elicit a truthful personification of the self, or activate a radical suspicion about its reality is, however, dependent on the modalities and forms of authentication legal knowledge chooses to deploy. The visible mediation of other agencies, most notably the nurse, meant for the Court that, regardless of his inner capacities, the display of Mauricio's self was ultimately dependent on the actions of a third party, and on that third party's ability to adequately interpret the signs produced. For the Court, this mediation triggered a moment of undecidability: Mauricio was so thoroughly embedded in systems of care and elicitation that the Court could not ascertain with full certainty a self-governing actor at the centre of the mesh of agencies, devices and processes of mediation with which it was presented. What was meant as a deed of personification – the production of 'today is Thursday' as evidence of Mauricio's intact will – was for the judges indistinguishable from an act of impersonation – of 'third persons' generating, controlling and interpreting his communications. The Court, in other words, chose to foreground the network and its connections at the expense of the hypothetical autonomous subject at its centre. It saw in the system of care and communication evidence of Mauricio's unknowability as a full-fledged legal person, rather than a machinery for rearticulating his intelligibility.

3 Technological enhancement, distributed personhood and the invisibility of mediation

Jose C.'s case offers an illuminating contrast. In the theatrical game of perspectives – between mask and person, foreground and background, role and impersonation – that defines the relationship between LIS patients and the law, Jose managed to appear in the eyes of the Court as a self-governing individual. In contrast to Mauricio's case, the socio-technical configuration through which he presented himself was, the Court concluded, a simple aid to the presentation of an intact self. Like Mauricio, Jose C. had used movements and an alphabet board to communicate with family and carers soon after the diagnosis of locked-in syndrome. Yet, after intense and prolonged physiotherapeutic care he was able to slowly regain full control over his right hand's index finger, and a degree of mobility in his neck. This brought about a radical change in the architecture of relations through which Jose could express himself and become intelligible to others as a conscious subject. With the help of an especially adapted trackball mouse, he was soon able to use an array of computer-based assistive technologies and, through them, regain some of his lost agential capacities. For example, voice reproduction software and a set of speakers attached to his wheelchair allowed him to produce speech through a digital voice. The Internet further extended these new communicative capacities by enabling him to interact and communicate with people outside his immediate environment via e-mail, social networking sites, blogs or chats, and was also instrumental in transforming intentions into actual and consequential actions, such as managing his bank accounts or purchasing goods and services online. Thanks to these computer-based interfaces, Jose wrote the first book-long autobiographical account of life with Locked-in Syndrome in Spanish, *El sindrome de cautiverio en zapatillas* (Carballo 2005).

Jose's case is indicative of the sorts of aids and mediations that are slowly being made available to LIS patients – aids and mediations that unsettle the legal assumption of the (healthy and able) biological body as the natural container of the person. By connecting the body with different combinations of hardware and software, such as wireless head-pointing devices, keyboard-scanning devices (which replicate on screen the *modus operandi* of the alphabet board), eye and gaze movement recognition interfaces, etc., the patient's agency is materially distributed along and enacted through novel socio-technical configurations. So-called 'environmental control units', for example, enable users to regain partial control over their home and work environments by operating different electronic appliances. Recent experimental developments in the field of neuroscience have taken the logic of these adaptive technologies a step further by employing brain–computer interfaces and deep-brain simulation to translate neural processes into outcomes without the use of muscles or further material connectors (Fenton and Alpert 2008; Schnakers *et al.* 2009). Although still experimental, these technologies have already produced some noteworthy breakthroughs. A group of researchers at the Wadsworth Center recently employed EEG technology, in which the subject wears a cap fitted with a series of electrodes connected to a computer, to translate the user's brain activity into simulated keystrokes and commands (Fenton and Alpert 2008: 123). Through this system, the patient can potentially learn to perform word processing, write e-mails, or move a robotic arm via the computer. Another research team recently implanted an electrode into the brain of a LIS patient, allowing him (through the use of software that translates brain signals into sounds) 'to produce three vowel sounds with good accuracy' (Smith 2005).

With these adaptive technologies the traditional mind–body interface is replaced by a complex socio-technical system as the means to express the self and to connect it with its surrounding environment. Rather than merely 'extending' or 'enhancing' the impaired capacities of a pre-existing person, these technologies constitute a scaffolding through which the LIS patient gains a new capacity to act, speak and be known as a full-fledged person. They can be seen as a radical example of Clark and Chalmers' (1998) thesis of the 'extended mind'. Challenging 'internalist' approaches that see cognitive processes as operations taking place within the brain, Clark and Chalmers advocate a form of 'active externalism' according to which certain cognitive processes can be seen as taking place in and through extended systems in which the brain is just one more element along other environmental and technological devices. One example of such 'extended' cognitive operation is the act of remembering, which rarely takes place 'in the brain' alone. We normally use all sorts of devices, from scribbled notes to PDAs, to inscribe, fix and retrieve our memories. According to this externalist view, these devices are not mere passive containers of our internal memories since they actively shape the form and extent of the cognitive processes and capacities through which we remember and, therefore, the very content of our memories – each technological configuration enables a particular form and structure of memory. Remembering, in other words, is not merely an operation that takes places through the neural operations of the brain: it requires the coupling of the brain and different environmental devices. 'If we remove the external component the system's behavioural competence will drop' (Clark and Chalmers 1998: 8–9). The brain and these external devices constitute a coupled system that can be considered a cognitive system in its own right.

The assistive technologies employed by LIS patients implode the difference between internal and external processes – not only for conventional 'cognitive' operations, but also for some of the agential and communicative capacities that commonly define personhood. Hardware and software devices are as functionally important as the brain in the production of a singularized and distinct person. These technologies complicate any easy demarcation between inside and outside, between the 'biological' individual and the 'technological' devices and processes on which it relies. The trackball mouse, voice-reproduction software, the wheelchair or the electrode cap constitute the architecture through which the LIS patient regains his capacity to act and become intelligible as a person. As the contrasting cases of Mauricio and Jose illustrate, different technological configurations enable different distributions of these agential and communicative capacities and, consequently, different forms of personhood. It is not just the mind, therefore, but the person itself that emerges as a distributed system – that is, as a coupling of the body and extended technological devices.[3]

Jose C.'s litigation affords us an opportunity to observe, in a striking moment of juridical redescription, how legal knowledge might apprehend such a distributed system as a fully formed *persona*, and how, in so doing, the legal system may be upholding a notion of citizenship compatible with the requirement of continuous, intense and all-encompassing care. As Jose recounts in his book (Carballo 2005), he decided to appeal the declaration of total incapacity when he realized that many of the elderly people living in his nursing care facility, individuals whose mental capabilities had visibly deteriorated, retained nevertheless their full legal rights, including the right to vote, whereas he had been deprived of full legal personhood despite the fact that his cognitive and volitional capacity remained intact following the onset of LIS. Against the opposition of the public prosecutor, the Appeals Court ruled in Jose's favour and mandated the 'reintegration' of his capacity. In all but a critical step the Court followed the reasoning that the Supreme Court applied to Mauricio's unsuccessful appeal. As in Mauricio's case, the medical assessments submitted to the Court confirmed Jose's 'full cognitive and volitional capacity', as well as the severe impairment of his physical abilities (Jdo. De Primera Instancia de Valladolid [2006], 00030/2006). What led the Court to reverse the declaration of total incapacity was Jose's demonstrated ability to 'materially carry out his decisions through the assistance of a computer with Internet' (ibid.). The Court noted in particular his capacity to manage his own bank accounts over the Internet, drawing once again the long-standing connection between the ability to manage oneself and the capacity to administer one's own property. 'The judicial examination', the Court noted,

> confirms his physical suffering, the conservation of his cognitive and volitional faculties, and his ability to use technical means to express his will, even to the point of carrying out on his own some of these decisions, all of it after being subject to a physiotherapeutic treatment that has allowed him to recover the degree of mobility necessary to make use of auxiliary means of communication.

The difference with respect to Mauricio's case rested on the availability in this instance of different technological platforms, particularly those facilitating access to the Internet. While Mauricio's display of personhood – via alphabet board, pen and nurse – was seen as the presentation of a fully intermediated subject, always at the mercy of the actors and devices through which it projected himself, the Appeals Court saw in Jose's actions on the Internet proof of his capacity of self-government, and of his relative independence from the actions of third parties. 'Technical means to express his will' and 'auxiliary means of communication' were also available to Mauricio, yet in that case these enhancements of the patient's powers of communication did not make him knowable to the law as a self-governing self; they rather triggered the Court's suspicion.

The reason for the divergent outcomes in Mauricio's and Jose's cases is to be found in what the Supreme Court called the 'factual reality of the person'; that is, in the way each of the respective socio-technical systems of care allowed each patient to evince his personhood. Although the alphabet board enabled Mauricio to communicate, he was dependent on the actions and mediations of a third person. This mediation prompted a moment of undecidability in which the Court could not know whether the *persona* that emerged from these mediations was a true representation of Mauricio's self or a simple act of impersonation. Rather than the expression of a self-sufficient person, the Court chose to see a form of ventriloquism that delegated to a third, unaccountable person the power to materialize Mauricio's thoughts and intentions. The introduction of assistive computer-based technologies radically altered the conditions of knowability of Jose's self. By adhering his body to different interfaces, Jose was able to bypass the mediation of any co-present third person, and to appear in the eyes of the Court as the indisputable source of the words and actions that emerged from this system. The redistribution and enhancement of his capacities did not result in a dispersal of his personhood, but rather in its intensification (see, in this context, Mialet 1999). That is, the redistribution enabled Jose to emerge as an actor demonstrably capable of thinking, communicating and acting autonomously. The severe physical and motor impairments became secondary, as the Court came to believe it could judge, and verify, his power of self-government by directly linking actions and words to mental states without the mediation of physical processes or the intervention of visible others.

The legal system was able to devolve to Jose full legal and political rights because the system composed of Jose's body and different assistive technologies produced the kind of evidence legal knowledge normally demands for the recognition of natural persons. The operations carried out on the Internet – the Court, as we noted, was especially taken by Jose's management of his bank accounts – were accepted as the indubitable expression of a self-governing actor. The discrepancy with the Supreme Court's ruling on Mauricio's case suggests that the nature of the components of the socio-technical system of care and mediation which LIS patients come to depend on makes a difference as far as their legal status *qua* political subjects is concerned. This raises an important concern over the equity of a system that grants – or in this case restores – full citizenship only to those patients who have access to expensive and technologically intensive forms of assistance, but that deploys the full scale of legal suspicion when those interfaces appear mundane or particularly laborious.

In the case of Jose C., the Court was able to balance the law's emphasis on 'self-government' as the fulcrum of full citizenship with the appreciation of the forms of intensive care and distributed action that characterize the form of life under Locked-in Syndrome. The restoration to Jose C. of his full civil *persona* offers an opportunity to rethink citizenship as a fragile *position* in need of constant care, rather than as an inalienable *condition* inscribed in our bodies. It is this notion of citizenship, a notion compatible with the reliance on intense and distributed forms of socio-technical support and knowledge, that we want to explore in the following section.

4 Citizenship as care

When it struck down the initial declaration of total incapacity, the Appeals Court suspended Jose's regime of *tutela* – tutelage by a legal guardian – and introduced in its place a different legal figure, that of *curatela*. *Curatela* harkens back to the *cura* of Roman law: 'a guardianship that protects the interest of youths … or incapacitated persons' who, while being recognized as *sui juris* (that is, possessing full legal and political rights), were in need of temporary or partial protection (Black's *Law Dictionary*, 7th edn, 1999: 386). The term is generally translated into English as 'conservatorship' or 'curatorship', but the Spanish word conveys etymological connections to 'care' and 'cure' that these English words barely express.[4]

With the change from *tutela* to *curatela*, the Appeals Court shifted the focus of the law's interest, from the transference and representation of Jose's will to the supervision of the forms of care to which he was entitled. The key distinction between the two conditions is that under *curatela* the actor retains – or, in Jose's case, recovers – full civil rights, including the right to vote and administer his property, whereas in a situation of *tutela* the subject is divested of his legal *persona*. The role of the court-appointed 'curator' is limited to assisting the person in relation to the physical needs specified by the Court; the purpose of the legal intervention is to supervise the provision of the forms of care and help that would enable a partially incapacitated individual to carry out the actions necessary for his sustenance (see Código Civil [2008], Title X, chs II, III). 'It is not necessary,' the Court ruled on Jose's appeal, 'to make up for[5] the will of the claimant, which he conserves in full, but to assist him in the material execution of those acts he chooses to do but is unable to carry out on his own.' Rather than replacing or representing the subject's will, then, the curator's function is 'to strengthen, control and channel' it (Jdo. De Primera Instancia de Valladolid [2006] 00030/2006).

The care received by a patient placed under the legal regime of *curatela* should in principle be no different from the care he would receive under the condition of *tutela*. The difference is that under *curatela* that care is all the law concerns itself with, whereas in a situation of *tutela* the law is primarily preoccupied with the handover of rights and the mechanics of legal mediation that follow a declaration of total incapacity. Thus *curatela* describes a legal complementing of the person that is squarely focused on the care of the body, its needs and the material execution of its desires, rather than on the legal representation of the will that the figure of the legal guardian implies. In this final section we would like to explore the implications of such a figure, a legal *persona* that is deemed complete (from the point of view of rights and entitlements), but at the same time is defined in terms of the physical body's lacks and needs. For what the Appeals Court effectively construed in the case of Jose C. was a model of citizenship that rests on an assortment of devices, interfaces and communicative prostheses, a legal subject that displays his powers of self-government through biotechnological interfaces, a *persona* that is the result of socio-material mediations but can be known by the law with the pristine clarity of the fully autonomous self. Unable to locate the source of personhood in the self-contained, healthy body, the Court nevertheless recognized Jose's distributed personhood as a legally able subject, disentangling his disability and physical dependency from the issue of his citizenship.

The case of the LIS patient before the law resonates with Annemarie Mol's (2008) discussion of the forms of care that accompany life with diabetes. Mol argues that we should understand the ailing body, the body in need of care, as a fully formed, even the standard form of political life. Mol wants to challenge those philosophies of citizenship that, explicitly or implicitly, are premised on the possession of a body that can be 'controlled, tamed or transcended' by the individual's will (2008: 31), a notion of the political subject for which reliance on the care of others – persons or things – mars the ideal of the self-governing and autonomous citizen. Would Mol's argument still hold at its limit: a conscious human body completely reliant on and enmeshed in practices of care, the individual afflicted by LIS? The Appeals Court's ruling on Jose's case, and particularly the legal figure of *curatela*, shows that it does, by describing a status that reconciles all-encompassing care and full citizenship.

Care of the LIS patient is, needless to say, much more critical, forceful and comprehensive than that received by individuals with diabetes. It involves supplying all the material and physical requirements of life: ensuring adequate oxygenation and preventing the complications caused by immobility and incontinence, in addition to providing assistance for all forms of physical activity – breathing, swallowing, positional changes, etc. The eyes, often the main instrument of communication, need to be protected against corneal ulceration; pathological crying, a condition common to LIS patients (Bauer *et al.* 1979), is sometimes treated with selective serotonin re-uptake inhibitors. The intensity of these activities demands a complex, and expensive, system of care that includes a full-time professional carer, the attention and effort of relatives, and a constellation of artefacts and technologies, such as feeding tubes, a multi-position bed, a wheelchair and other especially adapted vehicles to transport the patient.

And yet, despite the intensity of this care, the arguments put forward by Mol in her defence of *patientism* – the understanding that living with a disease can provide a new standard model of the citizen – can illuminate the legal and political position of LIS patients. First, because the body in care, even the body of a LIS patient, is an active body. 'In order to stay alive,' Mol writes, 'a body cannot just hang together casually. It has to act' (2008: 39). The relationship of the body to the practices of its care is not a passive one: 'We do not engage in care despite, but with, our bodies' (2008: 40). This is nowhere more evident than in patients with Locked-in Syndrome. The LIS-affected body must act, not only in the socio-technically mediated fashion we discussed earlier when describing the aids that enable communication, but in the very manner in which the body responds and adjusts to the practices of care and to its providers. The relationship of the LIS patient to his body cannot be described in terms of control – the traditional relationship between will, desires and physical actions has been thoroughly broken, and the body is fully reliant on a distributed system of life support – but the active participation of the patient is nevertheless part and parcel of the regime of care. Care does not simply happen to the patient; he remains engaged in the nursing and nourishing activities that, now distributed along an ever more complex set of devices, technologies and people, sustain his ability to be an actor in the world.

An element of this engagement is captured in medical discussions of the LIS patient's involvement in decisions about his care, in oft-heard arguments about the need to consider him a party in the assessment of treatment options. 'At the bedside,' write Plum and Posner in *The Diagnosis of Stupor and Coma* (4th edn 2007: 7), 'discussion should be *with* the patient, not, as with an unconscious individual, *about* the patient.' Smith and Delargy (2005: 407), reporting on their own experience caring for LIS patients, write that: 'Although cognitive ability should not be overestimated, survivors' views regarding the focus of acute treatment, rehabilitation goals, and life choices should be formally sought'. In the guidelines for the 'care and management of profoundly and irreversibly paralyzed patients with retained consciousness cognition', produced by the American Academy of Neurology in 1995 and which include, as 'the most extreme example', individuals afflicted by LIS, it is clearly stated that such patients should be in a position to make decisions about their treatment choices. The presence of consciousness is the critical factor: 'Clinical decision-making for these patients should proceed along the same line as clinical decision-making for non-paralyzed, competent patients, that is, physicians have the obligation to follow the health care decisions competently made by their patients' (American Academy of Neurology 1995).

Yet the activity and engagement of the LIS patient in his care goes beyond the provision of information and his ability to make punctual decisions about his care. It extends to the patient's relationship to the different components of the assemblage of people, artefacts and devices that sustain his personal and social existence. In her analysis of how the body afflicted by neuromuscular disease adjusts to the wheelchair, Winance noted the hard work on the material and emotional links between body and device necessary for the latter to become a personal prosthesis; it is only through 'hard and lengthy work' (2006: 66), on the part of the device, the patient, and those assisting their mutual adaptation, that the aid 'becomes part of the body (and the person) in the sense that it modifies the way the person perceives, moves, and relates to the world' (2006: 58–59). The 'common materiality' that Winance describes as the effect of this process of mutual adjustment is at the same time enabling and disabling – it defines the transitions between comfort and pain, it constitutes what is allowed and what is forbidden (Winance 2006: 66). What we want to suggest here is that the LIS patient carries out forms of work and engagement – cognitive, physical and emotional – that go well beyond the provision of 'informed consent' to treatment decisions. This work, whether we characterize it as 'adjustment' to the artefactual environment or 'attuned attentiveness' to the care practices of others (Mol *et al.* 2010: 15), is essential to sustaining the collective forms of action that characterize the LIS patient's involvement in the world. *Curatela*, 'curatorship', is a good legal articulation of this position, in its ability to recognize physical disabilities and emphasize the importance and intensity of care – for the provision of which it creates a system of judicial supervision – while recognizing the full legal and political personhood of the patient. In contrast to the exclusive preoccupation with *autonomy* that shapes the regime of tutelage, *curatela* directs the law towards the complementing of the patient's *agency* through different systems of care and support (see Willems (2002) for a discussion of the distinction between autonomy and agency).

The centrality of care to Jose's restored legal *persona* takes us away from the notion of citizenship as an inalienable condition deriving from inherent human capacities – such as the capacity to perform certain cognitive operations or to carry out actions autonomously – and moves us in the direction of citizenship as a precarious position in need of constant care. Yet, as we have argued, to care and to be cared for are not subjective dispositions but complex practices embedded in socio-technical systems of support and knowledge. The case of LIS provides a telling example of the ongoing, collective effort that is required to produce and, crucially, *sustain* the conditions of intelligibility required to become a full-blown citizen. When seen from the perspective of care, citizenship emerges, then, as a position that is carved out and made available collectively.

5 Conclusion: infrastructures of care and the production of citizenship

Hard cases make bad law, as the adage goes. Locked-in Syndrome is perhaps a condition too exceptional to draw far-reaching conclusions. It offers, however, a valuable test case to explore how the boundaries of personhood are negotiated in the face of new forms of techno-scientific enhancement. The development over the past decades of novel biotechnologies, neurocognitive and computational interfaces, and other, more mundane assistive devices, has given rise to hybrid and distributed forms of personhood that call into question the identification of the person with the biological individual. The application of some of these technologies to LIS patients reveals that the capacities and processes that have customarily defined the person – agency, intentionality, speech – need not be performed within the confines of the biological body, but may be enacted through extended systems of care and knowledge.

The disparity of outcomes in the two cases discussed in this chapter illuminates the difficulties of legal knowledge in recognizing and adjudicating legal personhood once the biological boundaries of the body can no longer be taken as the obvious marker of the autonomous self. The LIS patient projects a fragile, highly mediated, techno-socially distributed form of personhood, whose conversion into a legal *persona* raises a number of difficulties. It is as if the law were for a moment thrown back to a situation of undecidability, and was forced to build the distinction between 'person' (*persona*) and 'thing' (*res*) from scratch again. Yet, as Pottage (in Pottage and Mundy 2004: 5) reminds us, 'the problem is that humans are *neither* person *nor* thing, or simultaneously person *and* thing, so that law quite literally makes the difference'.

As Marcel Mauss famously wrote, in the Roman world 'the person is more than an organisational fact, more than a name or a right to assume a role and a ritual mask. It is a basic fact of law' (Mauss 1985 [1938]: 14). This basic fact – the constitution of the person in opposition to things and actions – needs to be reconstituted and upheld in the face of a multiplication of hybrids thrown up by modern techno-scientific knowledge. The law is called to adjudicate over these new mixtures to determine which socio-material combinations are entitled to the rights of 'natural persons'. As we have seen in the cases of Mauricio and Jose, this process of adjudication is uncertain and contentious. It depends on producing, by novel means, evidence of the sorts of qualities the law has long associated with the recognition of natural persons. A specially adapted mouse, computer-based assistive technologies and Internet-based banking were used, in Jose's case, to demonstrate to the Appeals Court an ability to govern oneself. The alphabet board, the pen, the assisting nurse and Mauricio's persistence were not enough to display this same quality to the Supreme Court, for they left the judges with the lingering suspicion that what they had witnessed could in fact be an act of impersonation, rather than personification.

Mauricio thus remained in the limbo of *tutela* – his will represented, for key legal and administrative purposes, by his mother acting in her capacity of legal guardian. Jose, on the other hand, was reintegrated into the political community as a subject endowed with full civil and legal rights through the figure of *curatela* – a form of existence that reconciles full citizenship with an intense regime of care. With the imposition of *curatela* the Court de facto recognized that Jose's intelligibility as a political subject, as a citizen, could not be simply found within his body, but was dependent on an extended system of care, on the continuous use of a distributed network of expert knowledges and technologies. In so doing it opened the door to a notion of citizenship in which the biological individual is not seen as the necessary correlate of the political subject.

LIS patients are not alone in pushing the boundaries of legal personification. Patients in vegetative states, human embryos or individuals with mental and physical disabilities or suffering from addictions are other examples of 'boundary subjects', whose status as persons – and the nature of their civil and political rights – is dependent on the varying configurations of different systems of knowledge and care. The apparent proliferation of these liminal forms of personhood is partly the result of new techno-scientific platforms, partly an effect of the law's continuous interrogation of its own categories. In any case it challenges the viability of a notion of citizenship that takes the self-contained, self-mastered biological body as the ultimate standard, in a world increasingly populated by thoroughly mediated forms of life.

In this chapter we have argued for the need to depart from the notion of citizenship as an inalienable condition of the bounded biological individual, and to treat it instead as a position defined by the intersection between legal forms of personification and infrastructures of support, knowledge and care. The LIS patient is not, in this respect, different from his fellow citizens, but the extraordinary laboriousness – physical, technical, emotional – involved in producing evidence of his personhood before the law lays bare the socio-technical conditions of citizenship. The cases of Jose C. and Mauricio illustrate how the intelligibility of the patient as a self-governing person is dependent on the kind of *evidence* produced by the different mediations and prostheses through which the individual presents himself. Yet as the diversity of fates of Jose C. and Mauricio attests, the scrutiny of the law falls unevenly on the evidence produced by different technologies: some socio-material arrangements stand a better chance to be considered *mere* conduits or aids for the expression of the autonomous self. The implication is that citizenship is unequally distributed; it is a function of the relation between the type of evidence produced by different techno-scientific systems of care and of the specific modalities and forms of authentication legal knowledge chooses to deploy in each case. If we consider citizenship as an abstract, immaterial quality, residing somewhere in the self – a self that can be found, moreover, within the boundaries of the body – we will tend to miss these differentials. A socio-material perspective on citizenship, one that regards citizenship as the result of distributed, collective efforts, as a position sustained by relationships of care, knowledge and assistance, is a first step towards confronting the inequity of its distribution in an era of proliferating techno-biological hybrids.

Notes

1 Three varieties of the syndrome are generally identified in the medical literature: classic, incomplete and total (Bauer *et al.* 1979). In its classic form, the patient has full consciousness and is only capable of vertical eye movement. Patients suffering from incomplete LIS preserve or manage to recover other types of voluntary movement, whereas in the case of total LIS the patient suffers 'total immobility and inability to communicate, with full consciousness' (Smith and Delargy 2005: 406). There are chronic and transient forms of the condition in each of these three categories.

2 Smart *et al.* (2008: 451–452) describe the slow and highly elaborate progression of the diagnosis for an initially comatose patient who recovered spontaneous eye movement.

> Although it was possible to elicit reproducible eye movements to command, the patient's fluctuations in arousal and persistent ocular bobbing notably compromised the consistency of the responses. It was not until the arousal disorder and ocular bobbing resolved that he was able to consistently follow eye-movement commands.

3 Alfred Gell (1998) used the notion 'distributed person' to refer to the different ways in which the self becomes 'articulated' through the persons and objects that bear the sign of its agency; the objects we produce are *literally* externalized parts of our selves, and act as 'indexes' of our agency. It is in this sense that it is possible to claim that an artist becomes 'distributed' in the artworks she produces, or that her works are indexes of a 'distributed' person. Our use of the phrase in this chapter is different from Gell's. We employ it to describe the different modes in which extended techno-scientific systems of support and care enable the self to act and become intelligible as a distinct person.

4 The term 'conservatorship' is used in Anglo-American law to describe situations where an organization or individual is placed under the limited or temporary control of an external actor. For instance, in 2008 the US financial institutions Fannie Mae and Freddie Mac were placed into the 'conservatorship' of the Federal Housing Finance Agency. Mentally ill or severely disabled individuals are also commonly placed under 'conservatorship'. This legal form has also been used in efforts to remove individuals, against their immediate will, from religious sects.

5 The Spanish verb *suplir* may be translated as both 'replace' and 'make up for'.

References

American Academy of Neurology. 1995. 'Position Statement: Certain Aspects of the Care and Management of Profoundly and Irreversibly Paralyzed Patients with Retained Consciousness and Cognition', Report of the American Academy of Neurology's Ethics and Humanities Subcommittee.

Bauby, Jean-Dominique. 1998. *The Diving Bell and the Butterfly: A Memoir of Life in Death*, 1st edn. New York: Vintage.

Bauer, G., Gerstenbrand, F. and Rumpl, E. 1979. Variables of the locked-in syndrome. *Journal of Neurology* 221, 77–91.

Cairns, H., Oldfield, R.C., Pennybacker, J.B. and Whitteridge, D. 1941. Akinetic mutism with an epidermoid cyst of the 3rd ventricle. *Brain* 64(4): 273.

Carballo, Clavero J.L. 2005. *El syndrome de cautiverio en zapatillas*. Rico Adrados.

Clark, A. and Chalmers, D. 1998. The extended mind. *Analysis* 58(1): 7.

Código Civil. 2008. Thomson Civitas.

Col, L., TVSP, M. and Gupta, P. 2005. Locked-in Syndrome – a case report. *Indian Journal of Anaesthesia* 144.

Dumas A. 1996 [1844] *The Count of Monte Cristo*. Harmondsworth: Penguin Books.

Fenton, A. and Alpert, S. 2008. Extending our view on using BCIs for Locked-in Syndrome. *Neuroethics* 1(2) (6): 119–132. doi:10.1007/s12152-008-9014-8.

Gell, A. 1998. *Art and Agency*. Oxford: Clarendon Press.

Laureys, S., Pellas, F., Van Eeckhout, P., Ghorbel, S., Schnakers, C., Perrin, F., Berré, J., Faymonville, M.E., Pantke, K.H., Damas, F., Lamy, M., Moonen, G. and Goldman, S. 2005. The locked-in syndrome: what is it like to be conscious but paralysed and voiceless? *Progress in Brain Research* 150: 495.

Leon-Carrion, J., Van Eeckhout, P., Domínguez Morales, M. and Pérez Santamaría, F.J. 2002. "The Locked-in Syndrome: a syndrome looking for therapy". *Brain Injury* 16(7): 571–582.

Lulé, D., Zickler, C., Häcker, S., Bruno, S.A., Demertzi, A., Pellas, F., Laurey, S. and Cübler, A. 2009. Life can be worth living in locked-in syndrome. *Progress in Brain Research* 339–351.

Mauss, M. 1985 [1938]. 'A category of the human mind: the notion of person; the notion of self', in Carrithers, M., Collins, S. and Lukes, S. (eds) *The Category of the Person: Anthropology, Philosophy, History*. Cambridge: Cambridge University Press.

Mialet, H. 1999. Do angels have bodies? Two stories about subjectivity in science: The cases of William X and Mister H. *Social Studies of Science* 551–581.

Mol, A. 2008. *The Logic of Care: Health and the Problem of Patient Choice,* 1st edn. London: Routledge.

Mol, A., Moser, I. and Pols, J. 2010. *Care in Practice: On Tinkering in Clinics, Homes and Farms.* Transcript Verlag.

Patterson, J.R. and Grabois, M. 1986. Locked-in syndrome: a review of 139 cases. *Stroke* 17(4): 758.

Plum, F. and Posner, J.B. 2007. *Diagnosis of Stupor and Coma* 4th edn. Oxford: Oxford University Press.

Pottage, A. and Mundy, M. 2004. 'Introduction', in *Law, Anthropology, and the Constitution of the Social: Making Persons and Things.* Cambridge: Cambridge University Press.

Rose, Nikolas S. 2007. *The Politics of Life Itself: Biomedicine, Power, and Subjectivity in the Twenty-first Century.* Princeton, NJ: Princeton University Press.

Schnakers, C., Perrin, F., Schabus, M., Hustinx, R., Majerus, S., Moonen, G., Boly, M., Vanhaudenhuyse, A., Bruno, M.-A. and Laureys, S. 2009. Detecting consciousness in a total locked-in syndrome: an active event-related paradigm. *Neurocase* 15(4) (2): 271–277.

Smart, C.M., Giacino, J.T., Cullen, T., Moreno, D.R., Hirsch, J., Schiff, N.D. and Gizzi, M. 2008. A case of locked-in syndrome complicated by central deafness. *Nature Clinical Practice Neurology* 4(8): 448–453.

Smith, E. 2005. Locked-in syndrome. *British Medical Journal,* 330 (7488): 406–409.

Smith, E. and Delargy, M. 2005. Locked-in syndrome. *British Medical Journal* 330 (7488): 406.

Strathern, M. 1992. *After Nature: English Kinship in the Late Twentieth Century.* Cambridge: Cambridge University Press.

Willems, D. (2002) "Managing one's body using self-management techniques: practising *autonomy*". In *Theoretical Medicine and Bioethics* 31(1): 23–38.

Winance, M. 2006. Trying out the wheelchair: the mutual shaping of people and devices through adjustment. *Science, Technology and Human Values* 31(1): 52.

Zola, É. 2008. *Thérèse Raquin.* Arc Manor LLC.

4 'Step inside: knowledge freely available'

The politics of (making) knowledge-objects

James Leach

Introduction

A large advertising sign hangs outside the new British Library building on Euston Road in London. It reads 'Step Inside. Knowledge Freely Available'.[1] A good slogan, but what does it imply about the way knowledge is thought of in contemporary society? Obviously the Library is a repository for a huge number of books, recordings, manuscripts and so forth. One would have to say that it is *these* that are freely available (and it is wonderful that they are of course). But in what sense are they *knowledge*? Or rather why it is that the advertisers decide to promise, by the emphasis of that term, something already a *value*, already more than the papers and inks themselves: something people can take away as 'knowledge'?

The theme of this chapter is a contemporary global politics that makes it important to call bound papers, objects, and other media that a library holds *knowledge*. The reference points are not libraries and their holdings specifically, but rather artistic practices from the UK, Indonesia and Melanesia, interdisciplinary research projects in the UK, and current intellectual property law. Clearly these are meant as examples of a wider phenomenon. My contention is that a trend that renders diverse objects, practices, effects, relationships, and forms of information into a single category – that of 'knowledge' – establishes the conditions for two further moves. Each has political implications. These moves are, first, a normative impetus for knowledge to take forms that make its ease of transmission paramount and often in the process prioritise narrow utility over wider effect. This in turn validates an impatience on the part of policy makers with complexity and dispute (Strathern 2007). Second, that the current image of knowledge as a detachable, circulating object sets up the possibility for a false scale of accounting in which comparative judgements about value are made to the detriment of recognising wider diverse, social benefits. This is most obvious in the current drive towards measuring 'impact', a particularly inappropriate register for arts and humanities research.

The impetus to view practices, relationships, performances, inscriptions, the emergence of particular and skilled persons and so forth as knowledge-producing activities with transactable object production as the aim of the endeavour suits the formulation of a certain political economy. I suggest that in this contemporary use, 'knowledge' has come to be a normative term denoting something that can be abstracted from the context of its production, and to carry value with it. We should ask ourselves what the effects of imagining there is something called knowledge are something that, if not always freely available (as in the Library's promise), is always available to move in transactions of the kind appropriate to commodities. As Strathern writes:

> One effect of the self-avowed knowledge economy has been to turn information into currency. Use value appears to depend on exchange value. Many certainly hold this view of scholarly knowledge. People openly state that there is no point in having such knowledge if one cannot communicate it, and they mean communicate it in the same form, that is, as knowledge. (Arguably, 'knowledge' is communicated as 'information,' but insofar as it is meant to be adding to someone else's knowledge, the terms can be hyphenated.)
>
> (Strathern 2004: 2)

My argument is not that knowledge is always and inevitably commodified – that it always has a price attached to it – but rather that the *form* in which diverse processes come to have recognised value in current regimes is through producing objects with analogous qualities to commodities. That is, objects that can be abstracted from their context of production and nevertheless carry the value of that production as an intrinsic element of the object itself. Knowledge as a fetish object, if you like.

In many contemporary situations, processes which create value by positioning persons and things in generative relations are judged narrowly dependent upon the 'knowledge' they produce. Looking across a range of ethnographic situations suggests we must widen the frame. All too often, policy and precedent focus on an object and its value to the detriment of the processes whereby wider social value is created. Thus universities are increasingly concerned with 'knowledge transfer', producing 'useable knowledge', while the protection of 'cultural knowledge' (Brown 2003) and intellectual property (Lessig 2004; Vaidhyanathan 2006) threaten to stifle creativity itself. A recurrent theme emerges. The emphasis for claims, for calculating recompense, and for describing value, locates value in objects produced, not in the processes of production. It is control over and access to those objects (and by this I do mean to include formulations and expressions) that concerns people.

My objective here is to highlight the work that calling vastly disparate things 'knowledge' does towards that objectification and formulation of value, primarily as object-value. I question what it means to call such diverse phenomena as cultural property, computer software, traditional arts, Papua New Guinean's use of plants, books, new technological processes (and so on) 'knowledge'. What effect is that move having on the social and political worlds in which these things come into being? One clear effect is that the outcomes of different social processes appear the 'same' across contexts, with ongoing implications for strategies to control them as resources, policy decisions with regard to the administration of institutions, and so forth. This needs to be opened up to scrutiny. My method is to suggest that the key to unlocking the problem lies in a conceptual move towards analysis of the relations in which persons and objects come to have their existence and effects.

I approach this from the perspective of having studied various claims and modes of ownership in relation to the realm of intellectual and cultural property. It is as well to be clear about this from the start. The focus on ownership does give what I have to say a particular slant. Thinking of ownership has taken me down the route of describing the claims people make over knowledge productions, and how those claims describe or build upon diverse ways of recognising value.

Arts, process, effects

The first example of translating social processes into knowledge-objects is from contemporary Indonesia, where I was fortunate enough to work with colleagues in 2005 to 2006 (Jaszi 2009), and particularly to collaborate with Lorraine Aragon (Aragon and Leach 2008). The research was with people the Indonesian state designates 'traditional artists' for the purpose of proposed legislation designed to protect cultural heritage. Given the basis of this legislation in Western intellectual property law,[2] the concept of authorship was central to our investigations. However, we were also concerned with the value traditional arts have to their practitioners, and the likely effects of the legislation with its assumptions of authorship and rights on their practices. In fact, the proposed legislation had a rather pressing element: it proposed the ownership by the state, in perpetuity, of any cultural expression without an identifiable author.[3] The intention was that the introduction of intellectual property laws would prevent the appropriation and distortion of valuable traditional arts by 'outsiders'. Using intellectual property law highlighted the rights of creators and authors, but disenfranchised those who could not make such claims.

Given this structure to the law, what should we make of it when traditional artists in Java and Bali stood in line to deny that they are the creators of the objects and performances by which they live? Or, when they say that the innovations they have introduced to their practices to make them more appealing and relevant to their audiences should not be viewed as emerging elements within the tradition? These are not innocent questions. For the logic of denying individual authorship for aspects of a tradition, while claiming that one's innovations are one's own and not subject to claims by others, fits closely with the current impetus in the arena of world trade negotiations and international bureaucracies. These organisations seek to offer protection to communal heritage, and/or to offer rights to individual creators, through historically specific, if now widespread formats (Strathern 2006; Vaidhyanathan 2006). Copyright law seems to be tailor made to protect the interests of innovators in the arts, while 'cultural property' advocates often see creating inventories of traditional material as the most promising way forward in assuring correct attribution to indigenous and subject populations (Daes 1997; Sedyawati 2005) and see Brown (1998).

The increasingly global application of intellectual and cultural property law is based on assumptions about the individual as a self-contained creative entity, and about artistic works – and by extension cultures – as potentially alienable and commercialisable assets which should be attached to these creators or their surrogates through legal rights. These assumptions were formalised first in the national laws of Europe and the United States, and then in statements of international institutions such as the World Intellectual Property Organization (WIPO) and the United Nations Educational, Scientific and Cultural Organization (hereafter UNESCO) (see UNESCO 1978, 1984, 2001, 2003). Concepts of intellectual property have been developed within the discourse of these multinational organisations into a vision of cultural patrimony without which emergent nations lose part of their 'personality' and thereby appear incomplete (Handler 1991; Harrison 2000). But it is important to question on a number of levels whether artistic communities, much less states, are properly analogous to individuals within whom creativity and identity are said to lie internally resident (Leach 2003a, 2003b); or more pertinently still for the current demonstration, whether creative works and their stylistic elements are best conceived of as isolable property (knowledge objects) properly subjected to formal legal ownership in this mode.

In Indonesia, both Hindu and Muslim traditional artists to whom we listened were reluctant to define themselves as their artwork's 'creator', or to say that their work will become part of their art tradition's future canon. Artists often comment that they are just 'followers' (*penyusul*) of their ancestral tradition and that the term 'creator' (*pencipta*) is applicable only to God. More than just a humble attitude or a theological dogma eliminating individual innovations, Indonesian artists' elaborating comments and actions entail a challenging vision for current moves to make such practices into objects that may be governed by laws applying to intellectual *property*. These were visions of what their art is, and what it does.

Indonesian artists whom we met repeatedly made claims, but these claims could only be taken as ownership claims over their creations in the most roundabout sense. The notion of cultural property has come to stand in international policy arenas for a variety of objects, places, and indeed practices, which may be attributed to a cultural or ethnic group. In this sense, it covers things which, although tangible or intangible 'objects' (the latter being made into objects through their expression), are not appropriate for alienation from that group. They are elements of their internal identity (UNESCO 2001, Preamble) and thus cultural property debates have a distinctly moral and ethical cast (Leach 2003b; see e.g. UNESCO 1978, 2001; Greenfield 1990). But in order to be viewed in this way, such items have to be existent and thereby tangible. One cannot own a distinctive form of creative practice, only the expressions of that practice. It is these that UNESCO, the prime movers in defining and developing the notion of cultural property, focus upon to the extent that they recommend that 'to ensure identification with a view to safeguarding, each State Party shall draw up, in a manner geared to its own situation, one or more inventories of the intangible cultural heritage present in its territory' (UNESCO 2003).

In this way, cultural property follows the logic of intellectual property as current US and UK legislation defines it: existent objects that demonstrate creative work, or innovation and added value. Social forms and the vitality of communal use are not protected. This particular way of defining not only what can, and cannot, be owned through an opposition of creativity and practice (Hallam and Ingold 2007) to the objects and forms which emerge also obviates the possibility of recognising alternative modes and outcomes of creative practice and value generation. In the ethnographic material we collected, a significant group of senior and successful traditional artists across a range of genres in Indonesia understood communication and creativity (coming from both knowing, and innovating upon disciplined practices) to be where value is generated. Their emphasis on the coherence and importance of tradition stemmed from a sense that to be in a position of knowing allows more creative engagement. This is an achievement of relational positioning. It cannot be pinned down to things already made.

UNESCO have recently come close to articulating a similar logic (2001),[4] yet their continued emphasis on preventing alienation and on repatriation makes it appear as though it is the objects themselves that allow creativity. Value still lies in objects, sites, or codifiable (that is, static) practices. In the cultural property rendering, claims people make over owning tradition are viewed as claims to objects in order to maintain their internal integrity and thus their possibility for entering into innovation and development with all their faculties intact, as it were.

Aragon and I suggested that artists were seeking to make claims over achievements not so much in the realm of material productions, but rather in achievements of relational positioning, vis-à-vis their human fellows (sponsors, hosts, colleagues, kin and audiences) as well as deity (Aragon and Leach 2008). Their physical art is not the key achievement. Rather, their work – as either material art or performance – is both the communicative sign and physical realisation of their social or relational accomplishment, and thus a sign of their power.[5]

Of course, objects have effects on and within social relations (Gell 1998). The difference that Aragon and I highlighted is between a focus on an object as the outcome of artistic endeavour, and on ongoing transformations in relationships. We pointed then to something at a more fundamental level of difference to the idea of knowledge as contained in and ownable as objects, one which is if anything more apparent in Melanesia, the site of other investigations into creativity (Leach 2004) and its relation to cultural property (Leach 2003b; Sykes 2001). The arts in these Melanesian contexts, just as in places we visited in Indonesia, tend to obviate the distinction between making, product and effect; between the process of making, and having an effect through the finished object that is made (Leach 2002). Yet this distinction is crucial to intellectual property law, as it amounts to the distinction between idea and expression, with the expression as that which can be protected. Under this logic, such protection is appropriate because it is the expression, not the idea or the process of making, which has the value (value creation in transaction determined by consumer market). However, in Melanesia and Indonesia, we saw that tradition is not objects, nor fixed rights of people over objects. Rather, it is abilities in relation to deity, predecessors, and others with whom one sits in relations of mutual obligation (Leach 2006), and through the whole recognition and engagement, the person themself emerges.

Knowledge and social effect

Recent (and not so recent) scholarship in the social studies of science suggests that it is not just in Indonesia and Melanesia that the value generated in the social processes around what we call 'knowledge production' is wider than the value that the knowledge has as an object, attached to an individual. That is, the processes of production are just as clearly examples of the emergence of certain persons and positions of power, hierarchy, influence, and so forth. Yet the very different emphasis on which aspects of the process create transactable value in Melanesia and Indonesia allows a clear critique to emerge of the way in which other kinds of value are created and retained in knowledge-based and knowledge-making relationships.

I pause for a moment here to put a little pressure on what we might mean by 'effect' in thinking about processes in which knowledge emerges. I draw on a formulation by Marilyn Strathern: a simple hierarchical classification for data, information and knowledge (Strathern 2005). Strathern suggests that data is what comes into the senses, it is unprocessed stuff. Information is that data organised in some way. Data made comprehensible, grouped according to some logic or other. But knowledge is more than information, it is data organised in a way that has an *effect*. To know something is to have to take it into account (Strathern 1992, 1999), often to have to act because of it, in the light of it, or around it (if only to consider it irrelevant).

I draw then on an anthropological understanding that knowledge is information that has an effect. Where and how does knowledge have its effect? As above, it is in social worlds that knowledge has its effects. New knowledge about a historical figure may change not only what books say about that person, but the way the time was understood, the presence or absence of that person's thought in others decisions, etc. Even for a lone scientist interacting wholly with the material world, the same is true. For what her investigation is aimed towards, how the work is supported, where it can and is recognised, the impetus to discover, and so forth, are all socially constituted. The effects of new information, even in my lone scientist example, are never directed primarily to the physical world, as we may like to imagine. Through technology, information/knowledge may come to have utility, but utility too is a socially defined value. Knowledge may be about the material world, but it is directed and made relevant by socially constituted values and interests. The effect is not limited to mechanical applications.

Indeed, crucially, much of the effect of knowledge production is on the person of the producer. That is, the effect of their engagements is apparent in changes in their status, their visibility in their discipline, or the wider academy, in senses of self, in ability to act and have an influence on others' behaviour. The fact that I want to define knowledge as information which is organised in a way that has an effect, does not mean I am arguing that all knowledge is useful, or usable, and certainly not that it ought to be. Utility and effect are very different things. The 'effects' we discerned as vital in Indonesia and Melanesia are *also* vital in arts and sciences closer to home. The consequences of investigation, or generating knowledge, are unpredictable. It is multiple and non-instrumental. Utility is a much narrower concept about a particular effect of an object on the material world.

There is a complexity then to the production of knowledge which involves changes in the producer, the context of production, potential utility and adoption by others (with ensuing debates over control and ownership). Effects upon things, effects upon other people, effects upon the producer. All such effects are dependent upon each other in a complex system of relations between objects, persons, skills, techniques, contexts for reception (Hirsch 2004) and so forth. It is this that we gloss as 'knowledge'. But having put it like that, is it any wonder that the outcomes from different processes produce not only different kinds of person, but different modes of communication, different elements to be transacted between parties? Knowing things never happens in a vacuum.

Having provided this frame, let me take the discussion forward by describing interdisciplinary projects which demonstrate that in the processs of 'knowledge production' many different effects are apparent, again vitally creating different persons and different kinds of effect.

Value generation in art and science collaborations

In recent years there have been a series of conscious innovations in government-influenced academic practice to encourage interdisciplinarity. For example, in 2003, the United Kingdom Arts and Humanities Research Board[6] and Arts Council England established an 'Arts and Science Research Fellowships' scheme with the aim to support collaborative research in arts and sciences. The application material drew on a report published by the Council for Science and Technology on the arts and humanities in relation to science and technology which concluded that

> the greatest challenges for UK society … are all ones in which the arts and humanities and science and technology need each other.… In the circumstances of modern society and the modern global economy, the concept of a distinct frontier between science and the arts and humanities is anachronistic … the relationships between the arts and humanities and science and technology need to be strengthened further.… Many of the most exciting areas of research lie between and across the boundaries of the traditionally defined disciplines.[7]

The Arts and Science Research Fellowship Scheme aimed then

> to support collaborative research specifically between the fields of the creative and performing arts and science and engineering which [were] likely to have a wider impact within the subject communities and beyond, as well as … seek[ing] to explore wider questions about whether and how art and science can mutually inform each other.

Running science up against art in the experiential way that this scheme did highlighted the conceptual distinctions and similarities between arts and sciences for the participants. They seemed to take as given certain characteristics of each. My analysis pursued an exploration of the way distinctions between art-as-knowledge making and science-as-knowledge making were constituted for the participants by their conjunction in the scheme.[8] It is worth summarising some of this investigation as it demonstrates the different modes and kinds of 'knowledge', and how the processes of its creation have a wider effect than captured in knowledge objects.

The Fellowships were shaped by assumptions that scientists work with entities that are external to themselves, while artists create their work from within themselves. This in turn may be linked to the recognition of two distinct kinds of material. As an observer, it was possible to see that there are effects of having 'the world' as on the one hand viewed as an external reality, ontologically independent of the perceiver prior to action, and on the other hand, the world as a social reality that all perceivers are responsible for creating. Those were effects in how the scientist as a person or the artist as a person could see themselves as connected to the outcomes of their labouring, to the knowledge objects they produced. The scientists who agreed to participate in this art and science collaboration did so (and said they did so) because they were interested in an opportunity to better 'align' their perception of themselves and of their work; that is, to make their individual and internal sense of self apparent in their professional outputs. What is interesting is how scientific knowledge making did not leave room for these aspects. I link this below with the notion of the *utility* of science in contrast to the perceived 'expressiveness' of art. Both succumb to the overall impetus to make knowledge objects. Yet the mode of making those objects is different, and has different effects on perceptions of their value, and on the person producing them. The artists were interested in engaging with the scientists in order to access a specific kind of material for their making processes, and not with making visible a sense of themselves as additional to, and necessary for, the particular objects they produced as 'art'.

The differences were perceived as necessary to science and art – with science working on an objective external reality which demanded an absence of subjectivity in the results. Objective reality demands an objective method of investigation. Thus the person of the scientist is 'purified' (Latour 1993) from the form in which their work appears. Artists, on the other hand, did not have to purify subjective perspective from their outputs: they were expected and valued as an integral part of the form of the object itself. Scientists saw themselves as involved in a highly technical process of revelation of what is not perceived as artifice itself, but is constituted as real in the social/cultural process of its emergence (Latour 1999), and in the claims that are possible in relation to it, whether those be legal or personal. The purification demanded by the context of claim making meant that scientists had less personal scope for influencing the output as people themselves. Artist's *outputs*, instead, remain associated much more closely with them as unique, individual persons. Scientists were thus represented as not being creative in a subjective sense, but as establishing relations between things already there. It is the reorganisation of things already there that creates something they can claim as knowledge. Scientists had to show that their knowledge was not a function of their subjectivity. In scientific authorship, the claim is an epistemological claim, a truth claim, and it is valued as such (Biagioli and Galison 2003). In artistic authorship there is also a claim to truth, but a subjective (or intersubjective) truth which may or may not communicate to others (have utility).

These different forms of knowledge and the aesthetic demands of each form meant that the place where collaboration and exchange were possible was in the realm of the personal for the scientists, and in the realm of the material for the artist. I suggest that the idea of commonality which made the scheme plausible in the first place as a collaboration which involved knowledges which could somehow be combined was made possible by precisely the contemporary notion of 'knowledge' as intangible objects which can be externalised from their producers, and which appear to carry their value despite this abstraction.[9] The emphasis is on production: that both arts and sciences *produce* things.

For those sponsoring the scheme, art and science provided an interesting counterpoint to each other. Combining them had the potential to offer more in the way of possibility than either on their own could. But as Strathern has pointed out (although in a different context altogether (Strathern 1988)), in order to act, people must be one thing or another. In any action involving knowledge which another is supposed to make use of or respond to, the actor must commit to one form of appearance. So what happened in the scheme was that the artists and scientists became like caricatures of themselves: the scientists found themselves deeply committed to their method of objectivity, while the artists were continually reiterating their need for individual understandings, subjective combinations of ideas and so forth.

Of course, what the Council for Science and Technology was attempting to set up was the possibility that science or art could extend the effectiveness of their actions and objects through including other kinds of knowledge form in their constitution. But that version of extension, including another knowledge object within a hybrid output, suits and reinforces the possibility that such knowledge is produced *as objects* as something extractable from the producers. As art objects or scientific discoveries, they would take on the status of an object that can be abstracted from their context of production, and carry their value elsewhere.

What is interesting is precisely the contrast of science and art as two ways of generating different types of knowledge objects and persons. The analysis shows that science is not the only form of knowledge making in Western societies and that other forms of producing knowledge, like art, matter politically insofar as they entail different modes of generating relationships between objects and persons.

A second example of interdisciplinary collaboration demonstrates exactly the process whereby social processes are narrowed to produce recognisable object outputs as knowledge. In this case, the process was intended and managed as a facilitation of complex collaboration precisely so that participants could justify their involvement by attachment to visible objects. The project was organised by an experienced arts researcher. It involved psychologists and neuroscientists and the innovative contemporary choreographer Wayne McGregor. Sessions and meetings took place in both scientists' laboratories and in the dance studio. McGregor had a clear aim in mind. His established method of working is to expose himself to a lot of stimulation, often through forays into various disciplines, and then make his dance pieces as a kind of reprocessing of his impressions and understandings. The collaboration with five very different psychologists was his research period for a piece before it was made. Each psychologist was given the opportunity to join in group discussions, observe dance material being made, and to talk with Wayne and his dance company in depth. Some used the time allocated to them in the overall structure to perform experiments: motion capture equipment was used by one pair to investigate motor control. Others asked dancers to perform movement tasks while also speaking, or reading, etc. to generate data on which parts of the brain control which kinds of activities.

The project was widely seen as a great success. There were multiple outputs, including the critically acclaimed dance work Ataxia and scientific papers (deLahunta and Shaw 2006).[10] The agenda was clearly to provide a collaborative space in which people felt comfortable to practise their own form of expertise – and to view the collaborative time together as producing a commonly constituted resource from which each party could then draw data and undertake analysis, or make dance, in their own sphere and through their own skills and techniques. So although there was a collaboration, and many outputs from the project, there was never a suggestion that there would be a common outcome, a single object or event which would encompass all of the participants and represent all their different expertise and input. People worked together to generate data through their interactions. This was then organised into information by particular disciplinary players. It was useful to other participants to see processes of its organisation in specific disciplines. But it was not until this information took specific forms appropriate to certain social spheres of recognition and reception (Hirsch 2004) that they became *knowledge.*

These processes had multiple effects through multiple forms of outcome, many of them valuable, but not recognisable as knowledge objects. There was the performance that demanded hard work of dancers and choreographer to produce not just the piece but themselves in relation to an audience, a body of critics, and to each other. A becoming not of knowledge object, but of a person who, through practice and skill, has an external effect on others *as* they are being constituted as that person in the gaze of those others. The wider collaboration encouraged immersion in unfamiliar forms of practice and action, without demanding that all the outputs were recognisable as 'knowledge objects'. It was actually a very similar process for the psychologists. Here, although what they produce (papers in scientific journals) are more 'knowledge-like', more transactable anyway as object forms in order for them to have effects, these outputs had to be tailored to very specific contexts of reception. It is that very specificity which makes the value of the endeavour.

So where did the narrowing of these generative processes to definable 'knowledge objects' become problematic? Ownership of knowledge and attribution relating to outcomes are often a source of tension in such collaborations. To follow my argument so far, and as I elaborate below:

knowledge has its effects in specific systems of social relations, among specific groups of people. Work that artists and scientists undertake together may well produce an artistic outcome that can be shown in dance venues. But that context, that arena of reception, or to be consistent with my language, that system of relationships in which effect is registered, is different for the scientist. The output in that form does not realise the same value for each party. This makes disputes over the control of value produced by collaborative work common.

In the Choreography and Cognition Project, where each party was expected to go off with the commonly made resource (data), and make what they would from it, such conflicts did not occur. As the organiser wrote: 'Relationships with chosen collaborators grew as the first three hour conversations turned into long-term commitments and dialogue gave rise to agreements on concepts, sharing of aims and objectives and acceptance of different goals and needs' (deLahunta 2006: 480). Choreography and Cognition was organised in order that the outcomes were always going to be within each party's realm of effect. An 'acceptance of different goals and needs' was stated at the outset. Thus there was never any suggestion of conflict over outcomes. It is in situations where a common outcome is desired or produced that such issues become more obviously problematic.

I hope it is by now clear that these conflicts arise when we see knowledge as an object which is context free, abstracted and discrete from the relations of its production and effect. As Mitchell and Latour (Latour 1987; Mitchell 2002) have both shown in different ways, science, and its claim to universality, exportability, and expertise independent of context, is the image for knowledge that exacerbates such problems for other kinds of practice, other kinds of 'knowledge'. It encourages a view of knowledge as transactable in a straightforward sense – as an object in the image of a commodity. Maybe it is the term *knowledge* itself then which is troublesome as, in current usage, it suggests that things produced in very different arenas, for very different purposes, for very different kinds of effect, are commensurate with one another.

Now I *have* made knowledge forms comparable to one another by following Strathern and suggesting that what we mean by knowledge is information organised to have effect. But I have also suggested that there are more or less radical disjunctures between the reasons that effects occur, and that many effects of 'knowledge' may not be intended, maybe by products of productive endeavour (or rather, that production is a by-product of relationships). Knowledge really is only knowledge when it has effects. And that depends on a metaphysic, on a series of assumptions and expectations about what effect will look like, what will be valuable, and so forth.

It is this last point, about value, that returns us to the notion of utility and the demand that all 'knowledge' take a specific transactable that is useful to others. What I have pointed out through the interdisciplinary examples is that effect does not mean utility, and use-ability by another kind of practitioner may well not be dependent upon utility in its expected sphere of reception. McGregor's successful dance Ataxia actually drew more from conversations with and observation of a woman who suffered from the condition than it did from the technical information the psychologists were able to provide about which bits of the brain were affected, but that surely does not make their knowledge useless. In general, a lesson here is that utility cannot be specified in advance. But the overall point is that when we call things which are really complex systems of persons, skills, contexts, objects, ideas, and so forth 'knowledge', we are in danger of making widely different productions and intended effects appear as if they were commensurate with one another.

In any production there is an objectification of social processes. Making all the things that come out of academic work or interdisciplinary collaboration into 'knowledge' can have the effect of reifying those productions as value in their own right. Knowledge seen as one kind of thing, easily exported from the context of its production and its effect, is to mistake the value of creative and relational processes for one possible aspect of them.

The philosophical roots of Euro-American intellectual property law are generally located in Locke's exposition of labour-based individual ownership rights, in conjunction with a vision of individual creative genius traced to eighteenth-century and early nineteenth-century Romantic authors (Jaszi and Woodmansee 2003). The Lockean view imagines the value of art, or any created work, as emanating from the individual, via labour, and entering the artwork through the mechanical process of its creation. Romantic authorship as a model suggests that individual genius transforms ordinary human experiences into extraordinary original art. The artwork, now a detached possession or event, is considered inanimate, perhaps representing, but not containing, the creativity of its producer. What is more, its source of value can be translated, through the notion of labour, into economic recompense. In this model, the artwork or performance may 'move' those in the audience, but its greater effects, or revelation of a deeper reality is, in the Kantian philosophical tradition, an interior experience, individual to each perceiver. The relation that is highlighted by such formulations is between artist and created object, and between perceiver and world beyond, not among makers, collaborators, audiences, perceivers, and creations as aspects of each other.[11]

In contrast to this logic, this chapter drew a critique from the analysis of Melanesian and Indonesian traditional artists including musicians, composers, dancers, textile designers and theatre performers who locate the primary value of their artistic activity within a set of human and cosmological relationships that are realised, or sometimes transformed, through artistic performances and works. The artists and scientists I described above are doing similar things. They are working within certain procedures and expectations to produce not just knowledge objects, but themselves as persons, the disciplines they work in as distinct and complementary, methods and techniques, connections and relationships. It is for this reason that knowledge is not simply transferred, nor should it be. In Indonesia, the artists' claim was over the work of relational positioning. That is why we get no strong statements over the ownership of the object created.

I am suggesting that we can learn something about what we currently call knowledge from thinking about value in a wider sense than tends to happen when it is the object produced which is of concern, not the process of production.

Conclusion

These case studies elaborate my statement that there is a commonly observable phenomenon across many contexts in which 'knowledge' is produced. It is a move that renders multiple values generated by complex social processes into simple and often commodifiable value located in objects, as if those objects retained their value shorn of the social relations in which they have effect. In other words, knowledge becomes a matter of economy. I am not naïve about this. I understand that transformations in the description of entities such as those currently covered by the term 'knowledge' mean that those entities can have different effects, and indeed can sustain the generation of social relations and values of different kinds. Capitalist knowledge economies are a form for social relations after all. However, what I have described here is a common series of transformations in which reifications of knowledge objects may clearly be seen to make transformations in the kinds of value that the social processes they are abstracted from generate, and that these transformations contribute to the wider emergence of 'knowledge economies' with their social audit practices, impact assessment for research and universities, state ownership, and bureaucratic control over cultural production and appropriation of traditional and indigenous knowledge. This serves certain interests. That is a matter of politics.

Acknowledgements

In various ways, the following people have assisted or given substance to this chapter. Lorraine Aragon, Lee Wilson, Bronac Ferran, Mario Biagioli and Scott deLahunta. I also thank the following funding bodies for supporting the research upon which the chapter is based: Social Science Research Council (and Joe Karaganis), Arts Council England, the Arts and Humanities Research Council, and The Leverhulme Trust.

Notes

1 January 2010.
2 In which the creator/author receives rights over the material expression while allowing that object to circulate.

3 Traditional and cultural practices would thus be owned by the state, not by the groups within the state who practise them.
4 Article 8 – Cultural goods and services: commodities of a unique kind.

> In the face of present-day economic and technological change, opening up vast prospects for creation and innovation, particular attention must be paid to the diversity of the supply of creative work, to due recognition of the rights of authors and artists and to the specificity of cultural goods and services which, as vectors of identity, values and meaning, must not be treated as mere commodities or consumer goods.

5 The idea that traditional arts in Indonesia, particularly in Java, are relational is not unexplored. See, in particular, Keeler's sophisticated ethnography of Javanese puppet theatre and its sometimes inattentive audiences (Keeler, W. 1987. *Javanese Shadow Plays, Javanese Selves.* Princeton, NJ: Princeton University Press).
6 AHRB – which was soon to become a full Research Council – the AHRC – in 2004.
7 Council for Science and Technology. 2001. *Imagination and Understanding. A Report on the Arts, Sciences and Humanities in Relation to Science and Technology.* UK Government/Department of Trade and Industry.
8 Leach, J. In prep. Constructing Aesthetics and Utility: Art, Science and the Purification of Knowledge.
9 See Leach. Forthcoming.
10 www.choreocog.net/.
11 Aragon and I believe that such generalisations about what is in reality a highly complex and contested series of philosophical positions is justifiable because our comment is upon the simplification, the rendering of complex realities as all following the same logic, that intellectual property law effects.

References

Aragon, L. and J. Leach. 2008. Arts and Owners: Intellectual Property Law and the Politics of Scale in Indonesian Arts. *American Ethnologist* 35, 607–631.

Biagioli, M. and P. Galison (eds). 2003. *Scientific Authorship: Credit And Intellectual Property in Science.* New York; London: Routledge.

Brown, M.F. 1998. Can Culture Be Copyrighted? *Current Anthropology* 39, 193–206.

——. 2003. *Who Owns Native Culture?* Cambridge, MA: Harvard University Press.

Council for Science and Technology. 2001. *Imagination and Understanding. A Report on the Arts, Sciences and Humanities in Relation to Science and Technology.* UK Government/Department of Trade and Industry.

Daes, E.-I. 1997. *Protection of the Heritage of Indigenous People.* (Human Rights Study Series). New York: United Nations.

deLahunta, S. 2006. Willing Conversations. The Process of Being Between. *Leonardo* 39, 479–481.

deLahunta, S. and N.Z. Shaw. 2006. Constructing Memories: Creation of the choreographic Resource. *Performance Research* 11, 53–62.

Gell, A. 1998. *Art and Agency.* Oxford: Oxford University Press.

Greenfield, J. 1990. *The Return of Cultural Treasures.* Cambridge: Cambridge University Press.

Hallam, E. and T. Ingold (eds). 2007. *Creativity and Cultural Improvisation.* Oxford: Berg.

Handler, R. 1991. Who Owns the Past? History, Cultural Property, and the Logic of Possessive Individualism. In *The Politics of Culture* (ed.) B. Williams. Washington, DC: Smithsonian Institution Press.

Harrison, S. 2000. From Prestige Goods to Legacies: Property and the Objectification of Culture in Melanesia. *Comparative Studies in Society and History* 42, 662–679.

Hirsch, E. 2004. Boundaries of Creation: The Work of Credibility in Science and Ceremony. In *Transactions and Creations. Property Debates and the Stimulus of Melanesia* (ed.) E. Hirsch and M. Strathern. Oxford: Berghahn Books.

Jaszi, P. 2009. *Indonesian Traditional Arts – Issues Articulated by Artists and Community Leaders and Possible Responses.* Washington, DC: American University.

Jaszi, P. and M. Woodmansee. 2003. Beyond Authorship. Refiguring Rights in Traditional Culture and Bioknowledge. In *Scientific Authorship. Credit and Intellectual Property in Science* (eds) M. Biagioli and P. Galison. London: Routledge.

Keeler, W. 1987. *Javanese Shadow Plays, Javanese Selves.* Princeton, NJ: Princeton University Press.

Latour, B. 1987. *Science in Action. How to Follow Scientists Through Society.* Cambridge, MA: Harvard University Press.

——. 1993. *We Have Never Been Modern.* Cambridge MA.: Harvard University Press.

——. 1999. *Pandora's Hope: Essays on the Reality of Science Studies.* Cambridge, MA: Harvard University Press.

Leach, J. 2002. Drum and Voice: Aesthetics and Social Process on the Rai Coast of Papua New Guinea. *Journal of the Royal Anthropological Institute* 8, 713–734.

——. 2003a. *Creative Land. Place and Procreation on the Rai Coast of Papua New Guinea.* Oxford and New York: Berghahn Books.

——. 2003b. Owning Creativity. Cultural Property and the Efficacy of Kastom on the Rai Coast of Papua New Guinea. *Journal of Material Culture* 8, 123–143.

——. 2004. Modes of Creativity. In *Transactions and Creations. Property Debates and the Stimulus of Melanesia* (eds) E. Hirsch and M. Strathern. Oxford and New York: Berghahn Books.

——. 2006. Out of Proportion? Anthropological Description of Power, Regeneration and Scale on the Rai Coast of PNG. In *Locating the Field. Space, Place and Context in Anthropology* (eds) S. Coleman and P. Collins. ASA Monograph. Oxford: Berg.

——. Forthcoming. Constructing Aesthetics and Utility: Art, Science and the Purification of Knowledge.

Lessig, L. 2004. *Free Culture: How Big Media Uses Technology and the Law to Lock Down Culture and Control Creativity.* New York: Penguin.

Mitchell, T. 2002. *Rule of Experts: Egypt, Techno-Politics, Modernity.* Berkeley: University of California Press.

Sedyawati, E. 2005. Senu Pertunjukan Tradisi dan Hak Cipta. *Jurnal Seni Pertunjukan Indonesia* 13, 49–54.

Strathern, M. 1988. *The Gender of the Gift. Problems with Women and Problems with Society in Melanesia.* Berkeley: University of California Press.

——. 1992. *After Nature. English Kinship in the Late Twentieth Century.* Cambridge: Cambridge University Press.

——. 1999. *Property Substance and Effect.* London: Athlone Press.

——. 2004. The Whole Person and its Artefacts. *Annual Review of Anthropology* 33, 1–19.

——. 2005. *Kinship, Law and the Unexpected. Relatives are Always a Surprise.* Cambridge: Cambridge University Press.

——. 2006. Intellectual Property and Rights: An Anthropological Perspective. In *Handbook of Material Culture* (ed.) W.K.C. Tilley, S. Kuchler, M. Rowlands and P. Spyer. London: Sage.

——. 2007. Useful Knowledge. The 2005 Isiaah Berlin Lecture. In *Proceedings of the British Academy.* Oxford: Oxford University Press/British Academy.

Sykes, K. (ed.). 2001. *Culture and Cultural Property in the New Guinea Islands Region: Seven Case Studies.* New Delhi: UBS Publishers Distributors.

United Nations Educational, Social and Cultural Organization (UNESCO). 1978. *Contribution to the Implementation of the International Convention on Economic, Social and Cultural Rights in Light of the Decisions of the Economic and Social Council and of the Human Rights Committee Executive Board.* UNESCO.

——. 1984. *Protection of Movable Cultural Property.* UNESCO.

——. 2001. *UNESCO Universal Declaration on Cultural Diversity. Adopted by the 31st Session of the General Conference of UNESCO, Paris 2 November 2001.* UNESCO.

——. 2003. *Convention for the Safeguarding of the Intangible Cultural Heritage, Paris 17 October 2003.* UNESCO.

Vaidhyanathan, S. 2006. Afterword: Critical information Studies. A Bibliographic Manifesto. *Cultural Studies* 20, 292–315.

5 Informal knowledge and its enablements

The role of the new technologies

Saskia Sassen[1]

The rapid proliferation of global computer-based networks and the growing digitization of knowledge, which allows it to circulate in those global networks, unsettle the standard meanings of knowledge. This in turn weakens the effectiveness of conventional framings for understanding what we mean by knowledge. It makes legible the particularity of the supposedly 'natural' or 'scientific' categories through which formal institutions organize 'their' knowledge – knowledge that has been defined as the aims of these institutions.

Network technologies have the potential to open up both the established categories of formalized knowledge and of the associated knowledge practices. Thereby the actual bodies of knowledge can then more easily exit or go beyond hierarchical institutionalized controls. One indication of this possibility is that global civil society organizations, including those who are poor and weak, can access some (not all!) of the same advanced datasets that were once the preserve of professional knowledge elites – whether these are data about the financial crisis, the latest scientific discoveries about the toxicity of widely used chemicals, or a whole range of other specialized subjects. These bodies of knowledge can get disassembled according to the criteria or needs of a far broader range of users. Key pieces of that knowledge can then navigate through a range of digital networks. This process itself contributes to strengthen the distributive potential of digital networks, thereby further enabling that disassembling.

The valence of these digital capabilities is highly variable – it can be good or bad for society, and all the in-between possibilities. Those who organize the annual sales of stolen credit card numbers benefit as much from these capabilities as does Forest Watch, with is complex networks bridging rainforest activists in the forests with their central offices in Washington, DC.

Here I examine these critical dimensions in order to understand how technologies with enormous distributive potential can be used with very different aims in mind. One dimension of this variability, and the one of concern in this chapter, is that they may be used to democratize but also to concentrate power. In principle, the range of empirical cases we could use to examine some of these issues is vast. I will examine the sharp differences between the use that each, high-finance and civil society organizations, makes of the distributive potential of these technologies – a potential that has strong connotations of democracy and participation. The specifics of each case allow us to identify particular patterns. One key fact that comes out of this type of analysis is that the logics of the users (including digitized actors) actually shape the technical outcome. They do so in diverse ways and to variable extents, as the comparison in this chapter will show. In this variability, the logics of users contribute to constitute what is knowledge, for whom it is useful, and who has access to it.

The question of knowledge in digital networks

What were once unitary bodies of knowledge ensconced in specific categories, often housed in closed institutions, can now get redeployed in bits and pieces across diverse institutional orders. This redeployment may occur through a variety of means and instruments. What differentiates the digital redeployment of knowledge from more traditional non-digital ones is the scope, speed, and the resulting multiplier effects of that redeployment, whether it is a closed or an open network.

This redeployment of what was once ensconced in closed formal categories has the effect of informalizing at least some features of that knowledge through their disassembling and reassembling into novel mixes. These novel mixes of bits and pieces are likely to be informal, at least initially. A second critical dimension is that these reassembled and informalized bodies of knowledge can feed into both novel and existing conditions – including political, economic, technical, cultural, and subjective conditions, which can be strengthened or weakened, democratized or not. Opening up established categories and informalizing particular components of formal knowledge may be seen as positive – for example, it can help democratize spheres once subject to hierarchical controls. Or it may be seen as negative – for example, the redeployment of particular financial regulations that has led to the so-called shadow banking system, which is, strictly speaking, legal but no regulator can control given the speed of transactions and interconnected markets.

Singling out processes that informalize existing categories and bodies of knowledge allows one to capture what are often highly dynamic but not particularly legible moments in a trajectory that may well wind up formalizing some of the informal. It allows us to capture a far broader range of instances than if we confine the focus to formalized knowledge and practices. Indeed, in my research I also find emergent informalities at the center of highly formalized systems (Sassen 2008: chs 7 and 8). This type of analysis goes well beyond the more familiar notions of the incapacity of national state regulatory frameworks to govern novel conditions.

Again, here I examine these critical dimensions in order to understand their variable articulations and how technologies with enormous distributive potential can be used to democratize or to concentrate power. A basic proposition is the importance of capturing the diversity and specificity of 'socio-digital formations' (Latham and Sassen 2005: intro.), and hence the possibility of whole new types of articulation between politics and knowledge. In principle, the range of empirical cases we could use to examine some of these issues is vast. The specifics of each case allow us to identify particular patterns. Different kinds of socio-digital formations make legible different ways in which this articulation between informalized knowledge and political enablements can be constituted.

The two cases used to develop the argument empirically are electronic financial networks and electronic activist networks. Both cases are part of global dynamics and both have been significantly shaped by the three properties of digital networks: decentralized access/distributed outcomes, simultaneity, and interconnectivity. But these technical properties have produced strikingly different outcomes in each case. In one case, these properties contribute to distributive outcomes: greater participation of local organizations in global networks. Thereby they help constitute transboundary public spheres or forms of globality centered in multiple localized types of struggles and agency. In the second case, these same properties have led to higher levels of control and concentration in the global capital market even though the power of these financial electronic networks rests on a kind of distributed power; that is, millions of investors distributed around the world and their millions of individual decisions.

In spite of their enormous differences, both are interactive domains. For analytical purposes I distinguish the technical capacities of digital networks from the more complex socio-digital formations that such interactive domains help constitute. Intervening mechanisms that may have little to do with the technology per se can reshape network outcomes, and, as I will show, are one of the sharp differences between the use that high-finance and civil society make of the distributive potential of these technologies (with its strong connotations of democracy and participation). The fact of this reshaping by the social logics of users and digitized actors carries implications for political practices, including governance and democratic participation.

Financial networks and civil society networks also illuminate an emergent problematic about the extent to which the combination of decentralized access and multiple choices will tend to produce power law distributions regardless of the social logics guiding users. Thus civil society organizations may well produce outcomes similar to finance in that a limited number of organizations (e.g., Oxfam, Amnesty International, Greenpeace, etc.) concentrate a disproportionate share of influence, visibility, and resources. One way of thinking about this is in terms of political formats (see, e.g., Dean *et al.* 2006; Lovink 2003). That is to say, civil society organizations have been subjected to constraints that force them into a format – akin to that of incorporated firms with conventional accountability requirements – that keeps them from using the new technologies in more radical ways (Sassen 2008: ch 7). Thus I would argue that finance succeeds in escaping conventional formats when two or more financial exchanges merge and thereby constitute a networked platform, allowing them to maximize the utilities of network technologies; this is a very different type from that of mergers and acquisitions, which creates a larger but still conventional corporate format. In this sense, I would argue that finance has been far ahead of civil society in the use of networked technologies. It has actually invented new formats to accommodate its use: multi-sited networked platforms, where each financial center is a node in the network. Civil society organizations have had many obstacles put in their way towards these types of networked arrangements. In many ways they have been forced to take the form of incorporated firms rather than networked platforms. There is, in my analysis, a political issue here that is yet another variable which contributes to produce diverse socio-digital formations even when based on similar network technologies.

The particularities of these two cases serve to address several larger research agendas now under way. They include specifying, among others, advancing our understanding of the actual socio-digital formations arising from these mixes of technology and interaction (Latham and Sassen 2005; Barry and Slater 2002; Bartlett 2007; Bennett and Entman 2001; Berman 2002; Howard and Jones 2004; Mansell and Silverstone 1998; Schuler 1996), the possible new forms of sociality such mixes may be engendering (see e.g., Castells 1996; Dutton 1999; Elmer 2004; Himanen 2004; Latham and Sassen 2005; Olesen 2005; Whittel 2001), the possible new forms of economic development and social justice struggles enabled by these technologies (Avgerou 2002; Credé and Mansell 1998; Gurstein 2000; Leizerov 2000; Mansell and Steinmueller 2002), and the consequences for state authority of digital networks that can override many traditional jurisdictions (Bauchner 2000; Drake and Williams III 2006; *Indiana Journal of Global Legal Studies* 1998; Johnson and Post 1996; Klein 2005; Rosenau and Singh 2002).

The condition of the Internet, including social networking, as a decentralized network of networks has fed strong notions about its built-in autonomy from state power and its capacity to enhance democracy from the bottom up via a strengthening of both market dynamics and access by civil society. In a context of multiple partial and specific changes linked to globalization, digitization has contributed to the ascendance and greater weight of subnational scales, such as the global city, and supranational scales, such as global markets, where previously the national scale was dominant. These rescalings do not always parallel existing formalizations of state authority. At its most general these developments raise questions about the regulatory capacities of states, and about their potential for undermining state authority as it has come to be constituted over the last century.

But there are conditionalities which not even these technologies can escape. Among these we might mention the social shaping of technology (see e.g., Bowker and Star 1999; Coleman 2004; Latour 1996; Lievrouw and Livingstone 2002; Mackenzie and Wajcman 1999; Seely Brown and Duguid 2002), the limits of what speed can add to an outcome (see e.g., Mackenzie with Elzen 1994; Sassen 2008: ch. 7), the role of politics in shaping communication (see e.g., Dean 2002; Howard 2006; Lovink 2002; Mansell and Silverstone 1998), the built-in stickiness of existing technical options (see e.g., Newman 2001; Shaw 2001; Woolgar 2002), and the segmentations within digital space (Koopmans 2004; Lessig 1996; Loader 1998; McChesney 2000; Monberg 1998; Sassen 1999; Schiller 1995).

Digital formations of the powerful and the powerless

The technical properties of electronic interactive domains deliver their utilities through complex ecologies that include non-technological variables, such as the social and the subjective, as well as the particular cultures of use of different actors. One synthetic image we can use is that these ecologies are partly shaped by the particular social logics embedded in diverse domains.[2] When we look at electronic interactive domains as ecologies rather than as a purely technical condition, we make conceptual and empirical room for informal knowledge and knowledge practices.

Electronic interactive domains are inherently distributive given their technical properties, but once we recognize that social logics are at work in such interactive domains it is not necessarily the case that those distributive outcomes will be present every time. In politics, this distributive potential has led commentators to say that these electronic networks push towards democratizing outcomes. Again, this is partly an empirical question – it depends on what social logics (i.e., political project) is driving that network. In another finding that goes against much commentary, I have found that the higher the speed and the interconnectedness of the network in global finance, the greater the importance of informal systems of trust and of cultures of technical interpretation; that is to say, the more advanced the technologies, the more the need for deeply social capabilities when these technologies are used in interactive domains (Sassen 2008: ch. 7).

Thus while digitization of instruments and markets was critical to the sharp growth in the value and power of the global capital market, this outcome was shaped by interests and logics that typically had little to do with digitization per se. This brings to the fore the extent to which digitized markets are embedded in complex institutional settings (see e.g., Knorr Cetina and Preda 2004; Mackenzie and Millo 2003; Pauly 2002; Sassen 1991/2001), cultural frames (Pryke and Allen 2000; Thrift 2005; Zaloom 2003; more generally see Bell 2001; Trend 2001) and even intersubjective dynamics (Fisher 2006). And while the raw power achieved by the capital markets through digitization also facilitated the institutionalizing of finance-dominated economic criteria in national policy, digitization per se could not have achieved this policy outcome – it took actual national institutional settings and actors (Helleiner 1999; Pauly 2002; Sassen 2008: ch 5; for cases beyond the financial markets see, e.g., Barfield *et al.* 2003; Waesche 2003).

In short, the supranational electronic market, which partly operates outside any government's exclusive jurisdiction, is only one of the spaces of global finance. The other type of space is one marked by the thick environments of actual financial centers, places where national laws continue to be operative, albeit often profoundly altered laws. These multiple territorial insertions of private economic electronic space entail a complex interaction with national law and state authority. The notion of 'global cities' captures this particular embeddedness of various forms of global hypermobile capital – including financial capital – in a network of well over forty financial centers across the world.[3] This embeddedness carries significant implications for theory and politics, specifically for the conditions through which governments and citizens can act on this new electronic world (see e.g., Bousquet and Wills 2003; Kamarck and Nye 2002; Roseneau and Singh 2002; Sassen 2008: chs 5, 8 and 9), though there are clearly limits (Dean *et al.* 2006; Olesen 2005; Robinson 2004).

Producing capital mobility takes capital fixity: state-of-the-art environments, well-housed talent, and conventional infrastructure – from highways to airports and railways. These are all partly place-bound conditions, even when the nature of their place boundedness differs from what it may have been a hundred years ago when place boundedness was far more likely to be a form of immobility. But digitization also brings with it an amplification of capacities that enable the liquefying of what is not liquid, thereby producing or raising the mobility of what we have customarily thought of as not mobile, or barely so. At its most extreme, this liquefying digitizes its object. Yet the hypermobility gained by an object through digitization is but one moment of a more complex condition.

In turn, much place boundedness is today increasingly – though not completely – inflected or inscribed by the hypermobility of some of its components, products, and outcomes. More than in the past, both fixity and mobility are located in a temporal frame where speed is ascendant and consequential. This type of fixity cannot be fully captured through a description confined to its material and locational features. The real estate industry illustrates some of these issues. Financial firms have invented instruments that liquefy real estate, thereby facilitating investment in real estate and its 'circulation' in global markets. Even though the physical remains part of what constitutes real estate, it has been transformed by the fact that it is represented by highly liquid instruments that can circulate in global markets. It may look the same, it may involve the same bricks and mortar, it may be new or old, but it is a transformed entity.[4]

Perhaps the opposite kind of articulation of law and territory from that of global finance is evident in a domain that has been equally transformed by digitization, but under radically different conditions. The key digital medium is the public access Internet, and the key actors are largely resource-poor organizations and individuals (for a range of instances see, e.g., Bennett 2003; Dahlgren 2001; Dutton 1999; Friedman 2005). This produces a specific kind of activism, one centered on multiple localities yet connected digitally at scales larger than the local, often reaching a global scale. As even small, resource-poor organizations and individuals can become participants in electronic networks, it signals the possibility of a sharp growth in cross-border politics by actors other than states (Khagram *et al.* 2002; Warkentin 2001). What is of interest here is that while these are poor and localized actors, in some ways they can partly bypass territorial state jurisdictions and, though local, they can begin to articulate with others worldwide and thereby constitute an incipient global commons.

We see here the formation of types of global politics that run through the specificities of localized concerns and struggles yet can be seen as expanding democratic participation beyond state boundaries. I regard these as non-cosmopolitan versions of global politics that in many ways raise questions about the relation of law to place that are the opposite of those raised by global finance.

From the perspective of state authority and territorial jurisdictions, the overall outcome may be described as a destabilizing of older formal hierarchies of scale and an emergence of not fully formalized new ones. Older hierarchies of scale, dating from the period that saw the ascendance of the nation-state, continue to operate. They are typically organized in terms of institutional level and territorial scope: from the international down to the national, the regional, the urban, and the local. But today's rescaling dynamics cut across institutional size and across the institutional encasements of territory produced by the formation of national states (Borja and Castells 1997; Graham 2003; Swyngedouw 1997; Taylor 2004;).

Electronic financial markets: making informal politics

Electronic financial markets are an interesting case because they are perhaps the most extreme example of how the digital might reveal itself to be indeed free of any spatial and, more concretely, territorial conditionalities. A growing scholarship examines the more extreme forms of this possibility, vis-à-vis both finance and other sectors (see, e.g., Geist 2003; *Indiana Journal of Global Legal Studies* 1998; Korbin 2001). The mix of speed, interconnectivity, and enhanced leverage evinced by electronic markets produces an image of global finance as hypermobile and placeless. Indeed, it is not easy to demonstrate that these markets are embedded in anything social, let alone concrete, as in cement.

The possibility of an almost purely technical domain autonomous from the social is further reinforced by the growing role played by academic financial economics in the invention of new derivatives, today the most widely used instrument. It has led to an increasingly influential notion that, if anything, these markets are embedded in academic financial economics. The latter has emerged since the 1980s as the shaper and legitimator, or the author and authorizer, of a new generation of derivatives (Barrett and Scott 2004; Callon 1998; Mackenzie 2002). Formal financial knowledge, epitomized by academic financial economics, is a key competitive resource in today's financial markets; work in that field thus also represents the 'fundamentals' of the market value of formal financial knowledge; that is, some of these instruments or models are more popular among investors than others.[5] Derivatives, in their many different modes, embody this knowledge and its market value.

But these technical capabilities, along with the growing complexity of instruments, actually generate a need for cultures of interpretation in the operation of these markets, cultures best produced and enacted in financial centers – that is, very territorial, complex, and thick environments. Thus, and perhaps ironically, as the technical and academic features of derivatives instruments and markets become stronger, these cultures become more significant in an interesting trade-off between technical capacities and cultural capacities (Sassen 2008: ch. 7). We can then use the need for these cultures of interpretation as an indicator of the limits of the academic embeddedness of derivatives and therewith recover the social architecture of derivatives trading markets. More specifically, it brings us back to the importance of financial centers – as distinct from financial 'markets' – as key, nested communities enabling the construction and functioning of such cultures of interpretation. The need for financial centers also, then, explains why the financial system needs a network of such centers (Budd 1995; Sassen 1991/2001). This need, in turn, carries implications for territorially bounded authority, and signals the formation of a specific type of territoriality, one marked by electronic networks and territorial insertions. Global cities are a more general, less narrowly technical instance of this same dynamic, including sectors other than finance. And beyond these types of formations there are other types of multi-sited global geographies – such as those binding Silicon Valley to Bangalore and kindred spaces (see generally Borja and Castells 1997; Corbridge *et al.* 1994; Graham 2003;Taylor 2003).

Yet alongside these territorial insertions that give national states some traction in regulating even the most global of financial markets (and other kinds of global firms and markets), the massive increases in values traded has given finance a good measure of power over national governments. This increase is probably one of the most significant outcomes of digitization in finance, with three of its capacities particularly critical. One is the digitizing of financial instruments. Computers have facilitated the development of these instruments and enabled their widespread use. Much of the complexity can be contained in the software, enabling users who may not fully grasp either the financial mathematics or the software algorithms involved. Further, when softwaring facilitates proprietary rights it also makes innovations more viable. Through innovations finance has raised the level of liquidity in the global capital market and increased the possibilities for liquefying forms of wealth hitherto considered non-liquid. The overall result has been a massive increase in the securitizing of previously untradeable assets, including various kinds of debt, and hence a massive increase in the overall volumes of global finance. Mediated through the specifics of contemporary finance and financial markets, digitization may then be seen as having contributed to a vast increase in the range of transactions.

Second, the distinctive features of digital networks can maximize the advantages of global market integration: simultaneous interconnected flows and decentralized access for investors and for exchanges in a growing number of countries. The key background factor here is that since the late 1980s countries have de- and re-regulated their economies to ensure cross-border convergence and the global integration of their financial centers. This non-digital condition amplified the new capabilities introduced by the digitization of markets and instruments.

Third, because finance is particularly about transactions rather than simply about flows of money, the technical properties of digital networks assume added meaning. Interconnectivity, simultaneity, decentralized access, and software instruments, all contribute to multiply the number of transactions, the length of transaction chains (that is, the distance between instrument and underlying assets), and thereby the number of participants. The overall outcome is a complex architecture of transactions that promote exponential growth in transactions and value.[6]

These three features of today's global market for capital are inextricably related to the new technologies. The difference they have made may be seen in two consequences. One is the multiplication of specialized global financial markets. It is not only a question of global markets for equities, bonds, futures, currencies, but also of the proliferation of enormously specialized global sub-markets for each of these. This proliferation is a function of increased complexity in the instruments, in turn made possible by digitization of both markets and instruments.

The second consequence is that the combination of these conditions has contributed to the distinctive position of the global capital market in relation to several other components of economic globalization. We can specify two major traits; one concerns orders of magnitude and the second the spatial organization of finance. In terms of the first, indicators are the actual monetary values involved and, though more difficult to measure, the growing weight of financial criteria in economic transactions, sometimes referred to as the financializing of the economy. Since 1980, the total stock of financial assets has increased three times faster than the aggregate gross domestic product (GDP) of the twenty-three highly developed countries that formed the Organization for Economic Cooperation and Development (OECD) for much of this period; and the volume of trading in currencies, bonds, and equities has increased about five times faster and now surpasses it by far. This aggregate GDP stood at about US$30 trillion in 2000 and US$36 trillion in 2004, while the worldwide value of internationally traded derivatives had reached over US$65 trillion in the late 1990s, a figure that rose to US$168 trillion in 2001, US$262 trillion in 2004, and US$640 trillion immediately before the financial crisis broke in September 2008. In contrast, the value of cross-border trade was US$15 trillion in 2007 and that of global foreign direct investment (FDI) stock, US$11 trillion in 2007 (BIS 2004; IMF 2005). Foreign exchange transactions were ten times as large as world trade in 1983, but (according to the triannual survey of the BIS) seventy times larger in both 1999 and 2003, and even higher in 2007, even though world trade also grew sharply over this period.

A second major set of issues about the transformative capacities of digitization has to do with the limits of technologically driven change, or, in other words, with the point at which this global electronic market for capital runs into the walls of its embeddedness in non-digital conditions. There are two distinct aspects here. One is the extent to which the global market for capital even though global and digital is actually embedded in multiple environments, some indeed global in scale, but others subnational, notably, the actual financial centers within which the exchanges are located (Mackenzie and Millo 2003). A second issue is the extent to which it remains concentrated in a limited number of the most powerful financial centers notwithstanding its character as a global electronic market and the growing number of 'national' financial centers that constitute it (GAWC 2005; Sassen 2008: ch. 5). The deregulation of finance could conceivably have led to wide geographic dispersal of this most electronic and global of markets.

The sharp concentration in leading financial markets may be illustrated with a few facts.[7] London, New York, Tokyo (notwithstanding a national economic recession), Paris, Frankfurt, and a few other cities regularly appear at the top *and* represent a large share of global transactions. This holds even after the 9/11 attacks in New York that destroyed the World Trade Center (though it was mostly not a financial complex) and damaged over fifty surrounding buildings, home to much financial activity. The level of damage was seen by many as a wake-up call to the vulnerabilities of sharp spatial centralization in a limited number of sites. London, Tokyo, New York, Paris (now consolidated with Amsterdam and Brussels as EuroNext), Hong Kong, and Frankfurt account for a major share of worldwide stock market capitalization. London, Frankfurt, and New York account for an enormous world share in the export of financial services.

London, New York, and Tokyo account for 58 percent of the foreign exchange market, one of the few truly global markets; together with Singapore, Hong Kong, Zurich, Geneva, Frankfurt, and Paris, they account for 85 percent in this, the most global of markets. These high levels of concentration do not preclude considerable activity in a large number of other markets, even though the latter may account for a small global share.

This trend towards consolidation in a few centers, even as the network of integrated financial centers expands globally, is also evident within countries. In the United States, for instance, New York concentrates the leading investment banks with only one other major international financial center in this enormous country: Chicago. Sydney and Toronto have equally gained power in continental sized countries, and have taken over functions and market share from what were once the major commercial centers, respectively Melbourne and Montreal. So have Sao Paulo and Bombay, which have gained share and functions from respectively Rio de Janeiro in Brazil, and New Delhi and Calcutta in India. These are all enormous countries and one might have thought that they could sustain multiple major financial centers, especially given their multi-polar urban system. It is not that secondary centers are not thriving, but rather that the leading centers have gained more rapidly and disproportionately from integration with global markets. This pattern is evident in many countries, including the leading economies of the world.

In brief, the private digital space of global finance intersects in at least two specific and often contradictory ways with the world of state authority and law. One is through the incorporation into national state policy of types of norms that reflect the operational logic of the global capital market rather than the national interest. The second is through the partial embeddedness of even the most digitized financial markets in actual financial centers, which partly returns global finance to the world of national governments although it does so under the umbrella of denationalized (that is, global-oriented) components of the state regulatory apparatus. Global digitized finance makes legible some of the complex and novel imbrications between law and territory, notably that there is not simply an overriding of national state authority even in the case of this most powerful of global actors. There is, rather, both the use of national authority for the implementation of regulations and laws that respond to the interests of global finance (with associated denationalizing of the pertinent state capacities involved), and the renewed weight of that authority through the ongoing need of the global financial system for financial centers.

These conditions raise a number of questions about the impact of this concentration of capital in global markets which allow for accelerated circulation in and out of countries. The global capital market now has the power to 'discipline' national governments; that is to say, to subject to financial criteria various monetary and fiscal policies that previously may have been subject to broader economic or social criteria. Does this trend alter the functioning of democratic governments? While the scholarly literature has not directly raised or addressed such questions, we can find more general responses, ranging from those who find that in the end the national state still exercises the ultimate authority in regulating finance (see e.g., Helleiner 1999; Pauly 2002), to those who see in the larger global economy an emergent power gaining at least partial ascendance over national states (Gill 1996; Panitch 1996).

Even the immobile and bearers of local knowledge can be part of global politics

Digital media are critical for place-centered activists focused on local issues that connect with other such groups around the world. This is cross-border political work centered on the fact that specific types of local issues recur in localities across the world.[8] These are politics which, unlike hacktivism (Denning 1999) and cyberwar (Der Derian 2001), are partly embedded in non-digital environments that shape, give meaning to, and to some extent constitute the event. These forms of activism contribute to an incipient unbundling of the exclusive authority, including symbolic authority, over territory and people we have long associated with the national state. This unbundling may well occur even when those involved are not necessarily problematizing the question of nationality or national identity; it can be a de facto unbundling of formal authority, one not predicated on a knowing rejection of the national.

None of this is historically new; yet there are two specific matters that signal the need for empirical and theoretical work on their ICT enabled form. One is that much of the conceptualization of the local in the social sciences has assumed physical or geographic proximity, and thereby a sharply defined territorial boundedness, with the associated implication of closure. The other, partly a consequence of the first, is a strong tendency to conceive of the local as part of a hierarchy of nested scales amounting to an institutionalized hierarchy, especially once there are national states. Even if these conceptualizations hold for most of what is the local today, the new ICTs are destabilizing these arrangements and invite a reconceptualization of the local able to accommodate instances that diverge from dominant patterns. Key among these current conditions are globalization and/or globality, as constitutive not only of cross-border institutional spaces but also of powerful imaginaries enabling aspirations to trans-boundary political practice even when the actors involved are basically localized and not mobile.

Computer-centered interactive technologies facilitate multiscalar transactions and simultaneous interconnectivity among those largely confined to a locality. They may be used to further develop old strategies (see e.g., Lannon 2002; Tsaliki 2002) and to develop new ways of organizing, notably electronic activism (Denning 1999; Rogers 2004; Smith 2001; Yang 2003). Internet media are the main type of ICT used, especially email, for organizations in the global south confined by little bandwidth and slow connections. To achieve the forms of globality that concern me in this chapter, it is important that there be a recognition of these technical constraints among major transnational organizations dealing with the global South: for instance, making text-only databases, with no visuals or HTML, no spreadsheets, and none of the other facilities that demand considerable bandwidth and fast connections (Electronic Frontier Foundation 2002; Pace and Panganiban 2002: 113).[9]

As has been widely recognized, new ICTs do not simply replace existing media techniques. The evidence is far from systematic and the object of study is continuously undergoing change, but we can basically identify two patterns. One is of no genuine need for these particular technologies given the nature of the organizing, or, at best, underutilization.[10] Another is creative utilization of the new ICTs along with older media to address the needs of particular communities, such as using the Internet to send audio files to be broadcast over loudspeakers to groups with no Internet connectivity, or that lack literacy. The M.S. Swaminathan Research Foundation in southern India has supported such work by setting up Village Knowledge Centers catering to populations that, even when illiterate, know exactly what types of information they need or want; for example, farmers and fishermen know the specific types of information they need at various times of the seasons. Amnesty International's International Secretariat has set up an infrastructure to collect electronic news feeds via satellite, which it then processes and redistributes to its staff work stations (Lebert 2003).

Use of these technologies has also contributed to forming new types of organizations and activism. Yang (2003) found that what were originally exclusively online discussions among groups and individuals in China concerned with the environment, evolved into active non-governmental organizations (NGOs). The diverse online hacktivisms examined by Denning (1999) are made up of mostly new types of activisms. Perhaps the most widely known case of how the Internet made a strategic difference, the Zapatista movement, became two organizational efforts; one a local rebellion in the mountains of Chiapas in Mexico, the other a transnational electronic civil society movement joined by multiple NGOs concerned with peace, trade, human rights, and other social justice struggles. The movement functioned through both the Internet and conventional media (Arquilla and Ronfeldt 2001; Cleaver 1998; Oleson 2005), putting pressure on the Mexican government. It shaped a new concept for civil organizing: multiple rhizomatically connected autonomous groups (Cleaver 1998).

Far less well known is that the local Zapatistas lacked an email infrastructure (Cleaver 1998) let alone collaborative workspaces on the Web. Messages had to be hand-carried, crossing military lines to bring them to others for uploading to the Internet; further, the solidarity networks themselves did not all have email, and sympathetic local communities often had problems with access (Mills 2002: 83). Yet Internet-based media did contribute enormously, in good part because of pre-existing social networks, a fact that is important in social movements initiatives (Khagram, Riker and Sikkink 2002) and in other contexts, including business (see Garcia 2002). Among the electronic networks involved, LaNeta played a crucial role in globalizing the struggle. LaNeta is a civil society network established with support from a San Francisco-based NGO, the Institute for Global Communication (IGC). In 1993 LaNeta became a member of the Association for Progressive Communications (APC) and began to function as a key connection between civil society organizations within and outside Mexico. A local movement in a remote part of the country transformed LaNeta into a transnational information hub.

All of this facilitates a new type of cross-border politics, deeply local yet intensely connected digitally. Activists can develop networks for circulating place-based information (about local environmental, housing, and political conditions) that can become part of their political work, and they can strategize around global conditions – the environment, growing poverty and unemployment worldwide, lack of accountability among multinationals, and so forth. While such political practices have long existed with other media and with other velocities, the new ICTs change the orders of magnitude, scope, and simultaneity of these efforts. This inscribes local political practice with new meanings and new potentialities. These dynamics are also at work in the constituting of global public spheres that may have little to do with specific political projects (Krause and Petro 2003; Sack 2005), though they do not always work along desired lines (Cederman and Kraus 2005).

Such multiscalar politics of the local can exit the nested scalings of national state systems (see e.g., Drainville 2005; Williamson, Alperovitz and Imbroscio 2002).[11] They can directly access other such local actors in the same country and city (Lovink and Riemens 2002), or across borders (Adams 1996). One Internet-based technology that reflects this possibility of escaping nested hierarchies of scale is the online workspace, often used for Internet-based collaboration (Bach and Stark 2005). Such a space can constitute a community of practice (Sharp 1997) or knowledge network (Creech and Willard 2001). An example of an online workspace is the Sustainable Development Communications Network (Kuntze *et al.* 2002) set up by a group of civil society organizations in 1998; it is a virtual, open, and collaborative organization to inform broader audiences about sustainable development, and to build members' capacities to use ICTs effectively. It has a trilingual Sustainable Development Gateway to integrate and showcase members' communication efforts. It contains links to thousands of member-contributed documents, a job bank, and mailing lists on sustainable development. It is one of several NGOs whose aim is to promote civil society collaboration through ICTs; others include the Association for Progressive Communications (APC), One World International, and Bellanet.

The types of political practice discussed here are not the cosmopolitan route to the global. They are global through the knowing multiplication of local practices. These are types of sociability and struggle deeply embedded in people's actions and activities. They also involve institution-building work with global scope that can come from localities and networks of localities with limited resources, and from informal social actors. Actors 'confined' by domestic roles can become actors in global networks without having to leave their work and roles in home communities. From being experienced as purely domestic and local, these 'domestic' settings become microenvironments on global circuits. They need not become cosmopolitan in this process; they may well remain domestic and particularistic in their orientation and continue to be engaged with their households, and local community struggles, and yet they are participating in emergent global politics. A community of practice can emerge that creates multiple lateral, horizontal communications, collaborations, solidarities, and supports.

Notes

1 Saskia Sassen (www.saskiasassen.com) is the Robert S. Lynd Professor of Sociology and Member, The Committee on Global Thought, at Columbia University. Her recent books are *Territory, Authority, Rights: From Medieval to Global Assemblages* (Princeton, NJ: Princeton University Press 2008), and *A Sociology of Globalization* (Norton 2007).
2 For a full development of these various issues see Sassen (2008: chs 7 and 8).
3 For instance, the growth of electronic network alliances among financial exchanges located in different cities makes legible that electronic markets are partly embedded in the concentrations of material resources and human talents of financial centers, because part of the purpose is to capture the specific advantages of each of the financial centers (Sassen 2008: ch. 7). Thus, such alliances are not about transcending the exchanges involved or merging everything into one exchange.
4 I use the term *imbrication* to capture this simultaneous interdependence and specificity of each the digital and the non-digital. They work on each other, but they do not produce hybridity in this process. Each maintains its distinct irreducible character (Sassen 2008: ch. 7).
5 The model designed for Long Term Capital Management (LTCM) was considered a significant and brilliant innovation. Others adopted similar arbitrage strategies, despite the fact that LTCM did its best to conceal its strategies (Mackenzie 2003). Mackenzie and Millo (2003) posit that two factors ensured the success of option pricing theory (Black-Scholes) in the Chicago Board Options Exchange. First, the markets gradually changed (e.g., alterations of Regulation T, the increasing acceptability of stock borrowing, and better communications) so that the assumptions of the model became increasingly realistic. Second, the spread of a particular technical culture of interpretation in the context of globalized economic processes gradually reduced barriers to the model's widespread use. The performativity of this model was not automatic but 'a contested, historically contingent outcome, ended by a historical event, the

crash of 1987' (Mackenzie 2003: 138). On the supervising of risk in financial markets see, e.g., Izquierdo (2001).

6 Elsewhere (Sassen 2008: chs 5 and 7) I have developed this thesis of finance today as being increasingly transaction-intensive and hence as raising the importance of financial centers today because they contain the capabilities for managing this transactivity precisely at a time when the latter assumes whole new features given digitization.

7 Among the main sources of data for the figures cited in this section are the International Bank for Settlements (Basle); International Monetary Fund (IMF) national accounts data; specialized trade publications such as Wall Street Journal's WorldScope, Morgan Stanley Capital International; *The Banker*, data listings in the *Financial Times* and in *The Economist*; and, especially for a focus on cities, the data produced by Technimetrics, Inc. (now part of Thomsons Financial, 1999).

8 This parallels cases where use of the Internet has allowed diasporas to be globally interconnected rather than confined to a one-to-one relationship with the country or region of origin (see, e.g., Glasius *et al.* 2002).

9 There are several organizations that work on adjusting to these constraints or providing adequate software and other facilities to disadvantaged NGOs. For instance, Bellanet (Bellanet 2002), a non-profit set up in 1995, helps such NGOs gain access to online information and with information dissemination to the South. To that end it has set up Web-to-email servers that can deliver Web pages by email to users confined to low bandwidth. It has developed multiple service lines. Bellanet's Open Development service line seeks to enable collaboration among NGOs through the use of open-source software, open content, and open standards; so it customized the Open Source PhP-Nuke software to set up an online collaborative space for the Medicinal Plants Network. Bellanet has adopted Open Content making all forms of content on its Web site freely available to the public; it supports the development of an open standard for project information (International Development Markup Language – IDML). Such open standards enable information sharing.

10 A study of the websites of international and national environmental NGOs in Finland, Britain, the Netherlands, Spain, and Greece (Tsaliki 2002: 102) concludes that the Internet is mainly useful for intra- and interorganizational collaboration and networking, mostly complementing existing media techniques for issue promotion and awareness raising.

11 The possibility of exiting or avoiding scale hierarchies by actors lacking power does not keep powerful actors from using the existence of different jurisdictional scales to their advantage (Morrill 1999); nor does it keep states from constraining local resistance through jurisdictional, administrative, and regulatory orders (Judd 1998).

References

Adams, P.C. (1996). 'Protest and the Scale Politics of Telecommunications'. *Political Geography*, 15(5): 419–441.

Arquilla, J. and Ronfeldt, D.F. (2001). *Networks and Netwars: The Future of Terror, Crime, and Militancy*. Santa Monica, CA: Rand.

Avgerou, C. (2002). *Information Systems and Global Diversity*. Oxford: Oxford University Press.

Bach, J. and Stark, D. (2005). 'Recombinant Technology and New Geographies of Association', in R. Latham and S. Sassen (eds), *Digital Formations: IT and New Architectures in the Global Realm*. Princeton, NJ: Princeton University Press, 37–53.

Barfield, C.E., Heiduk, G. and Welfens, P.J.J. (eds) (2003). *Internet, Economic Growth and Globalization: Perspectives on the New Economy in Europe, Japan and the USA*. New York: Springer.

Barlett, A. (2007) 'The City and the Self: The Emergence of New Political Subjects in London', in *Deciphering the Global: Its Spaces, Scales and Subjects,* edited by S. Sassen, New York and London: Routledge, pp. 221–243.

Barrett, M. and Scott, S. (2004). 'Electronic Trading and the Process of Globalization in Traditional Futures Exchanges: A Temporal Perspective'. *European Journal of Information Systems,* 13(1): 65–79.

Barry, A. and Slater, D. (2002). 'Introduction: The Technological Economy'. *Economy and Society,* 31(2): 175–193.

Bauchner, J.S. (2000). 'State Sovereignty and the Globalizing Effects of the Internet: A Case Study of the Privacy Debate'. *Brooklyn Journal of International Law,* 26(2): 689–722.

Bell, D. (2001). *An Introduction to Cybercultures.* London: Routledge.

Bellanet (2002). 'Report on Activities 2001–2002', http://home.bellanet.org, accessed March 18, 2006.

Bennett, W.L. (2003). 'Communicating Global Activism: Strengths and Vulnerabilities of Networked Politics'. *Information, Communication & Society,* 6(2): 143–168.

Bennett, W.L. and Entman, R.M. (eds) (2001). *Mediated Politics: Communication in the Future of Democracy.* Cambridge: Cambridge University Press.

Berman, P.S. (2002). 'The Globalization of Jurisdiction'. *University of Pennsylvania Law Review,* 151: 314–317.

BIS (Bank for International Settlements) (2004). *BIS Quarterly Review: International Banking and Financial Market Developments.* Basel: BIS Monetary and Economic Development.

Borja, J. and Castells, M. (1997). *The Local and the Global: Management of Cities in the Information Age.* London: Earthscan.

Bousquet, M. and Wills, K. (eds) (2003). *Web Authority: Online Domination and the Informatics of Resistance.* Boulder, CO: Alt-x Press.

Bowker, G.C. and Star, S.L. (1999). *Sorting Things Out: Classification and its Consequences.* Cambridge, MA: MIT Press.

Budd, L. (1995). 'Globalisation, Territory, and Strategic Alliances in Different Financial Centres Source'. *Urban Studies,* 32(2): 345–360.

Callon, M. (1998). *The Laws of the Markets.* Oxford: Blackwell.

Castells, M. (1996). *The Rise of the Network Society: The Information Age: Economy, Society and Culture, Volume 1.* Oxford: Blackwell.

Cederman, L.-E. and Kraus, P.A. (2005). 'Transnational Communications and the European Demos', in R. Latham and S. Sassen (eds), *Digital Formations: IT and New Architectures in the Global Realm.* Princeton, NJ: Princeton University Press, 283–311.

Cleaver, H. (1998). 'The Zapatista Effect: The Internet and the Rise of an Alternative Political Fabric'. *Journal of International Affairs,* 51(2): 621–640.

Coleman, G. (2004). 'The Political Agnosticism of Free and Open Source Software and the Inadvertent Politics of Contrast'. *Anthropological Quarterly,* 77(3): 507–519.

Corbridge, S., Thrift, N. and Martin, R. (eds) (1994). *Money, Power and Space.* Oxford: Blackwell.

Credé, A. and. Mansell, R.E. (1998). *Knowledge Societies ... in a Nutshell: Information Technology for Sustainable Development.* Ottawa: International Development Research Centre (IDRC).

Creech, H. and Willard, T. (2001). *Strategic Intentions: Managing Knowledge Networks for Sustainable Development.* Winnipeg: International Institute for Sustainable Development.

Dahlgren, P. (2001). 'The Public Sphere and the Net: Structure, Space, and Communication', in W.L. Bennett and R.M. Entman (eds), *Mediated Politics: Communication in the Future of Democracy.* Cambridge: Cambridge University Press, 33–55.

Dean, J. (2002). *Publicity's Secret: How Technoculture Capitalizes on Democracy.* Ithaca, NY: Cornell University Press.

Dean, J., Anderson, J.W. and Lovink, G. (2006) *Reformatting Politics: Information Technology and Global Civil Society.* London: Routledge.

Denning, D. (1999). *Information Warfare and Security.* New York: Addison-Wesley.

Der Derian, J. (2001). *Virtuous War: Mapping the Military-Industrial-Media-Entertainment Network.* Boulder, CO: Westview Press.

Drainville, A. (2005). *Contesting Globalization: Space and Place in the World Economy.* London: Routledge.

Drake, W.J. and Williams III, E.M. (2006). *Governing Global Electronic Networks: International Perspectives on Policy and Power.* Cambridge, MA: MIT Press.

Dutton, W.H. (ed.) (1999). *Society on the Line: Information Politics in the Digital Age.* Oxford: Oxford University Press.

Electronic Frontier Foundation (2002). 'Activist Training Manual', presented at the Ruckus Society Tech Toolbox Action Camp, June 24 to July 2.

Elmer, G. (2004). *Profiling Machines: Mapping the Personal Information Economy.* Cambridge, MA: MIT Press.

Fisher, M. (2006). 'Wall Street Women: Navigating Gendered Networks in the New Economy', in M. Fisher and G. Downey (eds), *Frontiers of Capital: Ethnographic Reflections on the New Economy.* Durham, NC: Duke University Press.

Fisher, M. and Downey, G. (eds) (2006). *Frontiers of Capital: Ethnographic Reflections on the New Economy.* Durham, NC: Duke University Press, pp. 209–236.

Friedman, E.J. (2005). 'The Reality of Virtual Reality: The Internet and Gender Equality Advocacy in Latin America'. *Latin American Politics and Society*, 47: 1–34.

Garcia, L. (2002). 'Architecture of Global Networking Technologies', in S. Sassen (ed.), *Global Networks, Linked Cities.* London: Routledge, 39–70.

GAWC (Globalization and World Cities Study Group and Network) (2005). www.lboro.ac.uk/gawc/, accessed March 18 2006.

Geist, M. (2003). 'Cyberlaw 2.0'. *Boston College Law Review*, 44: 323–358.

Gill, S. (1996). 'Globalization, Democratization, and the Politics of Indifference', in J. Mittelman (ed.), *Globalization: Critical Reflections.* Boulder, CO: Lynne Rienner Publishers, 205–228.

Glasius, M., Kaldor, M. and Anheier, J. (eds) (2002). *Global Civil Society Yearbook 2002.* Oxford: Oxford University Press.

Goldsmith, J. (1998). 'Against Cyberanarchy'. *University of Chicago Law Review*, 65: 1199–1250.

Graham, S. (ed.) (2003). *The Cybercities Reader.* London: Routledge.

Gurstein, M. (ed.) (2000). *Community Informatics: Enabling Communities with Information and Communication Technologies.* Hershey, PA: Idea Group.

Helleiner, E. (1999). 'Sovereignty, Territoriality and the Globalization of Finance', in D.A. Smith, D.J. Solinger and S. Topik (eds), *States and Sovereignty in the Global Economy.* London: Routledge, 138–157.

Himanen, P. (2001). *The Hacker Ethic and the Spirit of the Information Age.* New York: Random House.

Howard, P.N. (2006). *New Media Campaigns and the Managed Citizen.* New York: Cambridge University Press.

Howard, P.N. and Jones, S. (eds) (2004). *Society Online: The Internet in Context.* London: Sage.

Indiana Journal of Global Legal Studies (1998). 'Symposium: The Internet and the Sovereign State: The Role and Impact of Cyberspace on National and Global Governance', 5(2).

IMF (International Monetary Fund) (2005). 'International Financial Statistics'. Washington, DC: IMF.

Izquierdo, J.A. (2001). 'Reliability at Risk: The Supervision of Financial Models as a Case Study for Reflexive Economic Sociology'. *European Societies,* 3(1): 69–90.

Johnson, D. and Post, D. (1996). 'Law and Borders – The Rise of Law in Cyberspace'. *Stanford Law Review,* 48: 1367–1402.

Judd, D.R. (1998). 'The Case of the Missing Scales: A Commentary on Cox'. *Political Geography,* 17(1): 29–34.

Kamarck, E.C. and Nye, J.S. (eds) (2002). *Governance.Com: Democracy in the Information Age.* Washington, DC: Brookings Institution Press.

Khagram, S., Riker, J.V. and Sikkink, K. (eds) (2002). *Restructuring World Politics: Transnational Social Movements, Networks, and Norms.* Minneapolis, MN: University of Minnesota Press.

Klein, H. (2005). 'ICANN Reform: Establishing the Rule of Law', prepared for the World Summit on the Information Society (WSIS), www.ip3.gatech.edu/images/ICANN-Reform_Establishing-the-Rule-of-Law.pdf, accessed March 18 2006.

Knorr Cetina, K. and Preda, A. (eds) (2004). *The Sociology of Financial Markets.* Oxford: Oxford University Press.

Koopmans, R. (2004). 'Movements and Media: Selection Processes and Evolutionary Dynamics in the Public Sphere'. *Theory and Society,* 33(3–4): 367–391.

Korbin, S.J. (2001). 'Territoriality and the Governance of Cyberspace'. *Journal of International Business Studies,* 32(4): 687–704.

Krause, L. and Petro, P. (eds) (2003). *Global Cities: Cinema, Architecture, and Urbanism in a Digital Age.* New Brunswick, NJ, and London: Rutgers University Press.

Kuntze, M., Rottmann, S. and Symons, J. (2002). *Communications Strategies for World Bank and IMF-Watchers: New Tools for Networking and Collaboration.* London: Bretton Woods Project and Ethical Media, www.brettonwoodsproject.org/strategy/commosrpt.pdf, accessed March 18 2006.

Lannon, J. (2002). 'Technology and Ties that Bind: The Impact of the Internet on Non-Governmental Organizations Working to Combat Torture'. Unpublished MA thesis, University of Limerick.

Latham, R. and Sassen, S. (2005). 'Introduction. Digital Formations: Constructing an Object of Study', in R. Latham and S. Sassen (eds), *Digital Formations: IT and New Architectures in the Global Realm.* Princeton NJ: Princeton University Press, 1–34.

Latour, B. (1996). *Aramis or the Love of Technology.* Cambridge, MA: Harvard University Press.

Lebert, J. (2003). 'Writing Human Rights Activism: Amnesty International and the Challenges of Information and Communication Technologies', in M. McCaughey and M. Ayers (eds), *Cyberactivism: Online Activism in Theory and Practice.* London: Routledge, 209–232.

Leizerov, S. (2000). 'Privacy Advocacy Groups versus Intel: A Case Study of How Social Movements Are Tactically Using the Internet to Fight Corporations'. *Social Science Computer Review*, 18(4): 461–483.

Lessig, L. (1996). 'The Zones of Cyberspace'. *Stanford Law Review*, 48: 1403–1412.

Lievrouw, L.A. and Livingstone, S. (eds) (2002). *Handbook of New Media: Social Shaping and Consequences of ICTs*. London: Sage.

Loader, B. (ed.) (1998). *Cyberspace Divide: Equality, Agency, and Policy in the Information Age*. London: Routledge.

Lovink, G. (2002). *Dark Fiber: Tracking Critical Internet Culture*. Cambridge, MA: MIT Press.

—— (2003) *My First Recession: Critical Internet Culture in Transition*. Rotterdam: VP2/NAi Publishing.

Lovink, G. and Riemens, P. (2002). 'Digital City Amsterdam: Local Uses of Global Networks', in S. Sassen (ed.), *Global Networks/Linked Cities*. New York: Routledge.

Mackenzie, D. (2003). 'Long-Term Capital Management and the Sociology of Arbitrage'. *Economy and Society*, 32(3): 349–380.

Mackenzie, D. with Elzen, B. (1994). 'The Social Limits of Speed: The Development and Use of Supercomputers'. *IEEE Annals of the History of Computing*, 16(1): 46–61.

Mackenzie, D. and Wajcman, J. (1999). *The Social Shaping of Technology*. Milton Keynes: Open University Press.

Mackenzie, D. and Millo, Y. (2003). 'Constructing a Market, Performing Theory: The Historical Sociology of a Financial Derivatives Exchange'. *American Journal of Sociology*, 109(1): 107–145.

McChesney, R. (2000). *Rich Media, Poor Democracy*. New York: New Press.

Mansell, R. and Silverstone, R. (1998). *Communication by Design: The Politics of Information and Communication Technologies*. Oxford: Oxford University Press.

Mansell, R. and Steinmueller, W.E. (2002). *Mobilizing the Information Society: Strategies for Growth and Opportunity*. Oxford: Oxford University Press.

Mills, K. (2002). 'Cybernations: Identity, Self-Determination, Democracy, and the "Internet Effect" in the Emerging Information Order'. *Global Society*, 16(1): 69–87.

Monberg, J. (1998). 'Making the Public Count: A Comparative Case Study of Emergent Information Technology-Based Publics'. *Communication Theory*, 8(4): 426–454.

Morrill, R. (1999). 'Inequalities of Power, Costs and Benefits Across Geographic Scales: The Future Uses of the Hanford Reservation'. *Political Geography*, 18(1): 1–23.

Newman, J. (2001). 'Some Observations on the Semantics of Information'. *Information Systems Frontiers*, 3(2): 155–167.

Olesen, T. (2005). 'Transnational Publics: New Space of Social Movement Activism and the Problem of Long-Sightedness'. *Current Sociology*, 53(3): 419–440.

Pace, W.R. and Panganiban, R. (2002). 'The Power of Global Activist Networks: The Campaign for an International Criminal Court', in P.I. Hajnal (ed.), *Civil Society in the Information Age*. Aldershot: Ashgate, 109–126.

Panitch, L. (1996). 'Rethinking the Role of the State', in J. Mittleman (ed.), *Globalization: Critical Reflections*. Boulder, CO: Lynne Rienner Publishers, 83–111.

Pauly, L. (2002). 'Global Finance, Political Authority, and the Problem of Legitimation', in T.J. Biersteker and R.B. Hall (eds), *The Emergence of Private Authority and Global Governance*. Cambridge: Cambridge University Press, 76–90.

Pryke, M. and Allen, J. (2000). 'Monetized Time-Space: Derivatives – Money's "New Imaginary"?' *Economy and Society*, 29(2): 329–344.

Robinson, S. (2004). 'Towards a Neoapartheid System of Governance with IT Tools', SSRC IT & Governance Study Group. New York: SSRC, www.ssrc.org/programs/itic/publications/knowledge_report/memos/robinsonmemo4.pdf, accessed March 18 2006.

Rogers, R. (2004). *Information Politics on the Web*. Cambridge, MA: MIT Press.

Rosenau, J.N. and Singh, J.P. (eds) (2002). *Information Technologies and Global Politics: The Changing Scope of Power and Governance*. Albany, NY: State University of New York Press, 275–287.

Sack, W. (2005). 'Discourse, Architecture, and Very Large-scale Conversation', in R. Latham and S. Sassen (eds), *Digital Formations: IT and New Architectures in the Global Realm*. Princeton, NJ: Princeton University Press, 242–282.

Sassen, S. (1991/2001). *The Global City*. Princeton, NJ: Princeton University Press.

—— (1999). 'Digital Networks and Power', in M. Featherstone and S. Lash (eds) *Spaces of Culture: City, Nation, World*. London: Sage, 49–63.

—— (2008). *Territory, Authority, Rights: From Medieval to Global Assemblages*. Princeton NJ: Princeton University Press.

—— (2009) 'Mortgage Capital and its Particularities: A New Frontier for Global Finance.' *Journal of International Affairs*, 62(1): 187–212.

Schiller, H.I. (1995). *Information Inequality*. London: Routledge.

Schuler, D. (1996). *New Community Networks: Wired for Change*. Boston, MA: Addison-Wesley.

Seely Brown, J. and Duguid, P. (2002). *The Social Life of Information*. Cambridge, MA: Harvard Business School Press.

Sharp, J. (1997). 'Communities of Practice: A Review of the Literature', March 12, www.tfriend.com/cop-lit.htm, accessed March 18 2006.

Shaw, D. (2001). 'Playing the Links: Interactivity and Stickiness in .Com and "Not. Com" Web Sites'. *First Monday* 6: www.firstmonday.dk/issues/issue6_3/shaw, accessed March 18 2006.

Smith, P.J. (2001). 'The Impact of Globalization on Citizenship: Decline or Renaissance'. *Journal of Canadian Studies*, 36(1): 116–140.

Swyngedouw, E. (1997). 'Neither Global nor Local: "Globalization" and the Politics of Scale', in K.R. Cox (ed.) *Spaces of Globalization: Reasserting the Power of the Local*. New York: Guilford Press, 137–166.

Taylor, P.J. (2004). *World City Network: A Global Urban Analysis*. London: Routledge.

Thomsons Financials (1999). *1999 International Target Cities Report*. New York: Thomson Financial Investor Relations.

Thrift, N. (2005). *Knowing Capitalism*. Thousand Oaks. CA: Sage.

Trend, D. (ed.) (2001). *Reading Digital Culture*. Oxford: Blackwell.

Tsaliki, L. (2002). 'Online Forums and the Enlargement of the Public Space: Research Findings from a European Project'. *The Public*, 9(2): 95–112.

Waesche, N.M. (2003). *Internet Entrepreneurship in Europe: Venture Failure and the Timing of Telecommunications Reform*. Cheltenham: Edward Elgar.

Warkentin, C. (2001). *Reshaping World Politics: NGOs, the Internet, and Global Civil Society*. Lanham, MD: Rowman and Littlefield.

Whittel, A. (2001). 'Toward a Network Sociality'. *Theory, Culture & Society*, 18(6): 51–76.

Williamson, T., Alperovitz, G. and Imbroscio, D.L. (2002). *Making a Place for Community: Local Democracy in a Global Era*. London: Routledge.

Woolgar, S. (ed.) (2002). *Virtual Society? Technology, Cyberpole, Reality*. Oxford: Oxford University Press.

Yang, G. (2003). 'Weaving a Green Web: The Internet and Environmental Activism in China', *China Environment Series*, No. 6. Washington, DC: Woodrow Wilson International Centers for Scholars.

Zaloom, C. (2003). 'Ambiguous Numbers: Trading Technologies and Interpretation in Financial Markets'. *American Ethnologist*, 30(2): 258–272.

6 Secularisation and the politics of religious knowledge

Bryan S. Turner

Introduction: religion, knowledge and authority

Religious institutions, values and cosmologies played a central role in establishing the foundations for and production of knowledge in medieval and early modern societies. In part, this control over knowledge production was a function of the special status of religious functionaries in largely illiterate societies, the role of religious institutions in preserving languages (Greek, Latin and Arabic), and the role of religious beliefs in legitimating political power. The original language of revelation has subsequently played a dominant social and cultural role, especially in terms of religious missions. In Islam, the Arabic of the Qur'an is held to be untranslatable and hence Arab-speaking intellectuals in Muslim countries have significant prestige and authority. Religion has also for a great variety of different political regimes normally played a major role in legitimating power through its collective rituals and institutions. This relationship between state and religion can be very close – in what Max Weber called, for example, 'caesaropapism' – or it can be somewhat distant – in the majority of modern liberal democracies, especially in the constitutional monarchies. The parallel between God and people, king and subjects and husbands and their families formed the basis of patriarchal theories of power throughout the West (Schochet, 1975). But whether close or distant, the relationship appears to be resilient, despite the process of secularisation. Although there is a general impression, especially in the United Kingdom, that with secularisation religious issues and perspectives have been excluded from public debate, in recent history religion has continued to play a powerful political role in societies as different as Poland and the United States (Casanova, 1994; Zubrzycki, 2006).

Of course, this hegemony of religion in the public domain has been constantly challenged by rationalist, empirical or scientific beliefs throughout the history of Western Christendom. These challenges have not been confined to modern times. One example may be taken from the life of Roger Bacon (1214–1293) who argued that, although the Bible is the foundation of knowledge, we can employ reason in the service of human knowledge. Bacon was eventually imprisoned around 1277 by Jerome of Ascoli (the future Pope Nicholas IV) for his controversial views on curriculum reform, theory of language, and his views on natural philosophy, all of which were condemned as 'suspect novelties'. However, the contemporary challenge of science appears to be deeper and more widespread, because it often represents a significant challenge to both the epistemology and ontology of Christian orthodoxy. The modern foundations of religious knowledge have been shaken not only by the challenge of scientific knowledge from Charles Darwin onwards but by the rise of a global network society, the spread of secular higher education, the democratisation and decentralisation of religious knowledge, and the competition for authenticity and influence within what sociologists refer to as the 'spiritual market place' (Roof, 1993). However, from the perspective of the sociology of knowledge, these conflicts between science and religion are neither won nor lost simply on the merit of the arguments alone, since these conflicts also represent struggles between various elites and their institutions for power. In terms of the sociology of Pierre Bourdieu (1987), the politics of knowledge involves a struggle over social capital in the field of cultural life. Conflicts over religious knowledge have the same sociological characteristics as conflicts in the secular academic field over values and truth (Rey, 2007). What may often appear as obscure theological disputes over esoteric matters are often driven by complex conflicts between intellectual elites. In order to understand secularism and the transformation of the fortunes of religious knowledge, we must pay attention to the changing social relationships between intellectual elites, and changes in the social structure that make the traditional hierarchy between the priesthood and the laity largely redundant in more democratic times. The politics of religious knowledge ultimately raises questions about how modern social systems can be legitimated.

What has replaced this shared set of experiences is a more individualised and hybrid religiosity which is compatible with consumerism and popular culture. These symbolic universes also have different forms of authority which I will describe here as the contrast between hierarchical and horizontal authority. My argument may be seen as reflection on Ernest Gellner's account of the transition from hunter-gatherer to agrarian to industrial societies in his tripartite division between production, coercion and cognition in his *Plough, Sword and Book* (1988). This division provides us with a useful way of thinking about the power of a literate clergy and its monopoly of cognitive power. However, with the development of different social systems, the traditional sources of power – the technical means of production and coercion – were replaced by knowledge, and hence with the rise of post-industrial societies the sources of power shifted from the Church to the university (Bell, 1974). The slow disappearance of the mysterious (which I describe in this chapter as a transition from the ineffable to the effable) appears in fact to be associated with late modernity in which consumerism in the second half of the twentieth century became more influential than industrial production and with the influence of television in daily life the image often appears to be more potent than the concept.

All religions have been influenced by these secular pressures, and much attention has been given to the question of authority in Islam. For a variety of reasons (Volpi and Turner, 2007), Islam is faced with a crisis of authority that has clearly exposed the political struggle over knowledge. While Islam has received considerable academic attention in recent years, Christianity, more than any other world religion, appears to sit at the eye of the storm of secularisation. Furthermore, the position of papal authority within the Roman Catholic Church is probably unique since the majority of other religious traditions such as Judaism, Buddhism and Hinduism tend to have highly localised sources of authority. In this respect, the Jewish rabbinate, which gets its support from local lay audiences, is typical of such decentralised authority. Competition between rabbis has fuelled the disputatious nature of Judaism. However, alongside secularisation, we will have to consider globalisation with some degree of analytical care. In the course of this discussion, the authority of religious knowledge has been eroded with the decline in the power of religious institutions and beliefs. While religion has lost power in Western educational systems, 'religious nationalism' can, with globalisation, come to play a major role in political legitimacy, especially where there are conflicts over ethnicity and national identity. However, in everyday life religion becomes continuously enmeshed in secular consumer culture.

Secularisation and religious responses

Religious institutions have responded to secularism through various strategies, but in Christianity and Islam revivalism and reformism have been characteristic movements to reverse the apparently ineluctable spread of secular values on the back of global consumerism. These reformist movements in Christianity and Islam have similar dimensions and in general terms may be described as the 'pietisation' of modern religious consciousness (Tong and Turner, 2008). The growth of female piety in Asia and the Middle East is associated with the emergence of an urban middle class, the spread of higher education and the emergence of low fertility among women. As a result, these pious women set up their own Quranic study groups and occasionally dispute the interpretations of the Qur'an offered by their mullahs and imams. Some of these women have emerged as public intellectuals – such as Fatima Mernissi – whose alternative interpretations of Islamic teaching on women and sexuality have become globally influential (Mernissi, 2003). Western feminists have of course continued to condemn veiling as a manifestation of arbitrary patriarchal power (Lazreg, 2009), but more ethnographically sensitive research suggests that female piety can also express female subjectivity and empowerment (Mahmood, 2005).

Alongside these reformist or fundamentalist developments in Islam and Christianity, there is also a surge in popular religions or 'spirituality' – such as New Age religions – that often assume an overtly consumerist character. Modern spirituality is typically non-institutional, individualistic and hybrid in both belief and practice. These two developments – pietisation and popular spirituality – raise interesting questions about the basis of religious knowledge and its politics in modern societies. These new forms of religion specifically raise questions about the adequacy, for example, of Max Weber's theory of authority in terms of tradition, charisma and legal rational modes, because the authority of messages on the Internet is not easily covered by Weber's conceptual framework. However, his general framework for the sociological analysis of religion remains in some respects congenial to my approach. Weber's contrast between the religion of the masses (that is geared to bringing wealth and delivering health) and the religion of the virtuosi (that is geared to generating and conserving meaning) lies at the heart of much contemporary discussion of whether religion (of the elite) can withstand the encroachment of the religion of the masses in a consumer society. The recent furore over, for example, Mel Gibson's *The Passion of the Christ* or Dan Brown's *The Da Vinci Code* may be taken as evidence of the threat to official orthodoxies from popular expressions of religion in Hollywood and in pulp fiction that have a mass appeal. Such manifestations of religion in popular culture well illustrate the threat of modern consumption to elites and their defence of literate and articulate orthodox belief. It could be held that popular or folk representations of religion have always been problematic, but these contemporary forms have an immediate, mass and global impact, and these popular interpretations cannot be easily controlled or supervised by powerful institutions such as the state, church or mosque.

If we are, in terms of authority systems, in a post-Weberian world, we may also be entering a post-Durkheimian society, where institutionalised religious hierarchies are no longer the driving force behind increasingly individuated expressions of religion. While these private expressions of spirituality flourish, more institutionalised or formal patterns of religion are eroded. The social solidarity and effervescence that Durkheim (2001) claimed was the social foundation of all classification of knowledge, that is, the systems of 'primitive classification' dividing the world into the sacred and the profane, may have broken down. Orthodoxy is increasingly becoming epistemologically irrelevant, and the collective structures of authorities that ultimately found their justification in these constructions of religious knowledge are becoming obsolete and are pushed to one side by new media of communication.

Perhaps the conservative decades of Catholicism under the guidance of the Pope John Paul II may be an exception to the general rule in a context where orthodoxy was vigorously defended by a reinvigorated Curia.

However, informed analysis of the condition of the Catholic Church after the death of the Pope suggests that he failed to address, and did not staunch, the serious division between liberals and conservatives over such issues as women in the priesthood or sexual mores. The current crisis of Western Christendom – such as the tensions in 2009 between Roman Catholicism and the Anglican Church over male priesthood – as a whole is precisely a crisis of who has authority to speak on matters that are crucial to its survival.

Post-secular society

In this chapter, special attention is given to Christianity and secularism rather than to any comparative analysis of religions. There are two reasons for this attention. The first is that Christianity appears to be closely, perhaps necessarily, tied to secularism (Smith, 2008). In 1793, Kant's *Religion within the Boundaries of Mere Reason* (1998) set the stage for this argument. Distinguishing between a 'reflecting faith' which follows the moral law out of rational conviction and cultic religion which attempts to influence God by rituals, sacrifice and magic, Kant saw a close relationship between Enlightenment and pietism. For Kant, the Enlightenment involved throwing off childish dependence and establishing one's intellectual and moral autonomy, just as pietism had thrown off religious dependency in order that the spiritual life of the individual could flourish. In this respect, Protestant Christianity as a reflective faith ultimately required no religious – or specifically ecclesiastical – props and it was therefore a secularising force. Weber (2002) on his return from his sojourn among the Baptist sects in the United States took up this theme in *The Protestant Ethic and the Spirit of Capitalism* when he argued that the unintended consequence of the salvation drive of the Protestant sects was no less than the disenchantment of the world. The possibility of secularism appears paradoxically to lie at the heart of the Christian quest to confront the world.

The second reason is that the philosophy of religion has enjoyed in the past fifty years or so an extraordinary vitality, and much of that vitality has been associated with attempts to understand the social, cultural and intellectual consequences of secularism. One may refer here to a number of recent influential works – *Religion* (Derrida and Vattimo, 1998), *The Religious Significance of Atheism* (MacIntyre and Ricoeur, 1969), *The Secular Age* (Taylor, 2007), and *The Future of Religion* (Zabala, 2005). A new concept – 'post-secular society' – has emerged out of these philosophical discussions and the direction of this debate has been somewhat guided by Jürgen Habermas (2002 and 2006). These philosophical currents have also found an echo in the revitalisation of the sociology of religion in the work of Robert Bellah, Thomas Luckmann, Niklas Luhmann, Roland Robertson and others, which has been documented in the recent spate of textbooks and handbooks. These philosophical and sociological debates have often come to the striking conclusion that secularism is not wholly alien or inimical to the Christian religion. The decline of institutional religious life leaves open a number of possible developments – spirituality, implicit religion and New Age religions. The idea which was popular in the 1950s and 1960s that modernisation meant the death of religion and the end of the authority of religious knowledge has been replaced by more subtle and complex interpretations. Just as modernisation is no longer treated as a uniform process, so secularisation is studied in its multiple forms.

The important issue here is not simply the revival of a branch of academic learning. What is at issue is the very nature of the social itself. Sociologists from Durkheim onward have been interested in religion, because it is assumed to contain the seeds of social life as such. This insight into the social importance of religion was the real point of Durkheim's critique of the individualist, rationalist theories of religious belief as they had been developed by Herbert Spencer, E.B. Tylor and Max Müller. For Durkheim, there were no 'false religions' and hence religion would not simply disappear before the onslaught of positivist science. The religious and the social are bound together, and hence the real question behind the secularisation thesis is not so much the 'death of religion' as 'the death of the social'. This is the real substance of post-secularism, namely, the question: Can a society function without a collective identity and ultimately some underlying notion of the sacred? The dominance of individualism and the privatisation of many public institutions during the heyday of economic neoliberalism have eroded public space and collective values. The spiritual becomes a mirror of the commercial, and hence modern religiosity is mobile, hybrid, subjective and passive. The modern spiritual marketplace of modern youth cultures is a 'mix-and-match' spirituality that is parallel to the consumer marketplace (Roof, 1993, 1999).

The death of God: emotivism and the rise of consumer society

Modern religious thought has struggled to come to terms with the erosion of religious authority – a struggle that was announced by Nietzsche's infamous prophecy that God is dead. In his Godless-theology, he prepared the way for Weber's rationalisation thesis. Nietzsche bemoaned the growth of an economic system in which material prosperity would lead eventually to spiritual impoverishment, and in *The Will to Power* (Nietzsche, 1968: 866), he attacked 'the economic optimism' which naively believes that increasing expenditure will increase human welfare when on the contrary 'man is diminished'. In his recent analysis of religion in a post-secular context, Habermas has also observed that the growth of material wealth has weakened the need for transcendence. One might add that prosperity and longevity have made the promise of Heaven less relevant or urgent in modern societies in which the number of centenarians has grown exponentially (Turner, 2009a).

If we look more closely at the history of theology from Ludwig Feuerbach (1762–1814) to Ernst Troeltsch (1865–1923) and beyond, we can see one movement, which is broadly speaking liberal, and which sought to develop the Christian message to make it relevant to an industrial secular society, and another movement which is evangelical and opposed liberalism in asserting the literal truth of the Bible and the mysterious ineffable character of the human encounter with the Person of Jesus Christ. Secular theology, as it came to be called, in making the Christian message intelligible in a modern industrial environment has the consequence of making the sacred less remote, mysterious and formidable. It is a movement away from what Rudolf Otto in 1917 called 'the numinous' in his *The Idea of the Holy* (1959). Although the numinous (the sacred or the holy) cannot be ultimately defined, it is experienced as the wholly Other or Sublime in which humans have a sense of creaturely dependence and awe. These experiences may be summarised in the idea of the *mysterium tremendum*. For Otto, all religions worthy of the name have the numinous at their core. These experiences presuppose a hierarchically organised world in which the sacred is in some sense above the profane, just as the prophet is above the priest, and the clergy are above the laity, and in which the divine message descends in a patriarchal structure to Man, thereby retaining its mysterious ineffable qualities. The whole trend of modern secular theology – and modern philosophy – has been to 'demythologise' the message of the Bible, making its message audible and intelligible in terms of our human concerns.

However, the long march towards popular, horizontal, effable religion is not a smooth, unchallenged, intellectual development. There are needless to say significant theological criticisms of the subjectivism of modern religious belief and practice. Karl Barth's monumental thirteen-volume *Church Dogmatics* (1969–1967) was an attempt to block off the slide towards theological liberalism and to assert the otherness and heteronomy of God's Word and Revelation against what he saw as the pervasive subjectivism and anthropomorphism of liberal theology. In his understanding of the human subject as an object posited by God, Barth constructed a hierarchical system of religious communication in which there is the inner world of the Trinity, God's relationship with humanity through Christ, the relationship of Jesus to humanity and finally humanity's own communication. Although this hierarchy does not overtly deny human autonomy, it emphatically affirms the priority of God's acts of Revelation (Macken, 1990). In this way, God's Revelation in history never becomes the possession of human beings (McCormack, 1995). We might reasonably say that Barth's theology attempted to draw a line between what is properly effable and what is ineffable in order to moderate the impact of Protestant liberalism. The tendency of Protestant theology at the time may be illustrated from the popular sermons and addresses of Friedrich Schleiermacher (1996) which assumed a romantic, affective and emotional hue (Mackintosh, 1937).

The growing strength of subjectivity in modern spirituality suggests that Barth's attempt to close down that avenue has failed at least at the popular level. Modern spirituality is communicative, audible and effable. Above all it is grounded in how individuals *feel* about the world. This problem also lies at the base of Alasdair MacIntyre's critique of secular modernity in terms of the rise of what he calls 'emotivism'. In *After Virtue* (MacIntyre, 1984: 12), MacIntyre complained that emotivism implies that 'all moral judgments are *nothing* but expressions of preference, expressions of attitude or feeling'. One can detect a parallel between a consumer society in which individual preferences are the driving force behind economic growth and hence a society in which the consumer is sovereign. Such an individual's preferences are irrational, emotional 'wants' since in a condition of 'post-scarcity' all previous 'needs' have become simply 'wants'. We may regard this development in moral philosophy as a reflection of the rise of consumerism and note a similar transition in religious life from a 'theology of unhappiness' in which a miserable life on Earth finds its compensation in Heaven and a 'theology of happiness' in which an affluent life finds its satisfaction in this world (Turner, 2009b). To survive these changes, religion has to become popular and must be followed closely by a popular theology. These changes in knowledge were matched by the gradual erosion of clerical power from the mid-nineteenth century onwards as the reform of universities involved the decline of ecclesiastical patronage, the growth of secular public universities and eventually the rise of the science park.

In post-war theology and especially in Germany, there were important developments in theology which embraced secularisation and sought to redefine religion to make it more relevant to modern society. Two figures stand out as important in this development – Rudolf Bultmann (1884–1976) and Dietrich Bonhoeffer (1906–1945). Bultmann tried to create a position between the Protestant fundamentalism of Karl Barth and the liberals by questioning the historical accuracy and relevance of the biblical record and the fact of the event of Jesus Christ in human history. For example, Bultmann argued that for a modern person the idea of Christ ascending into Heaven no longer made sense, since our cosmology does not have God and Heaven as it were spatially above the Earth. For the Christian message to make any sense it had to be demythologised (Bultmann, 1984). Although Bonhoeffer never developed a systematic theology on the same scale as Bultmann, he was also instrumental in promoting the idea of 'religionless Christianity' and a 'world without God'. Barth, Bultmann and Bonhoeffer were responding, although in very different ways, to the obvious decline in religious observance among the working class and to the equally obvious appeal of socialism to the alienated worker, and subsequently responding to the devastation caused by two world wars. If the churches faced the problem of the alienated, secular worker, enticed by the prospect of socialism, in the high watermark of capitalism (from about 1850 to 1950), the modern churches are faced with the issue of the indifferent, isolated consumer, enticed by the prospect of material happiness.

In the United Kingdom, these continental developments which drew much of their inspiration from German theology were rather slow in having an impact, but the debate eventually became notoriously centred around the figure of John Robinson (1919–1983), the Bishop of Woolwich, who promoted the death-of-God theology. He argued, following Bultmann, that traditional notions of God as a Person out there or up there were metaphors that required radical transformation if the Christian message was to have any relevance, and in *Honest to God* (1963) he canvassed the idea of God as the 'ground of all being'.

Recent philosophy of religion, especially around Rorty and Vattimo, has sought to tease out the full implications for theology and society at large of living in a post-God world. As we have seen, atheism and theism at least in a Christian context are somehow inseparable, and hence secularism is so to speak no stranger to Christianity. This argument was powerfully developed by MacIntyre (MacIntyre and Ricoeur, 1969) in *The Religious Significance of Atheism* in which he asserted ironically that in the modern world, being a theist is not a significant option since a strong version of atheism was no longer available. This type of argument was taken up by Rorty, who as a pragmatist was sceptical about the imperialist claims of rationalism and promoted instead the idea of 'weak thought'. Rorty's proposals appear to be liberal, pragmatic and modest. He claimed that 'religion is unobjectionable as long as it privatised – as long as ecclesiastical institutions do not attempt to rally the faithful behind political proposals, and as long as believers and unbelievers agree to follow a policy of live and let live' (Rorty 2005: 33). These modest liberal proposals are based on the more radical view that, since thought is weak, secularisation means that all questions about the nature of God are useless, because no ultimately credible answers are possible. Vattimo's position is not dissimilar, except that he derives secularisation more directly from Christian teaching. He argues, concentrating on the theme of charity in 1 Corinthians 13, that the Incarnation teaches us that God's sacrifice was to hand over his power and authority to humanity. Secularisation as the empowerment of human beings to live life to the full is the message of the Gospels – the truth shall make us free (Vattimo, 2005: 48).

Both Rorty and Vattimo concluded that the Catholic Church's use of literalism as a defence against science and as the foundation of a universal message required an objectivist metaphysics which 'weak thought' has rendered obsolete. This strategy results in serious difficulties when, for example, the Church denies the priesthood to women on the grounds that it is not their 'natural' vocation (Vattimo, 2005: 49). This objectivist metaphysics also makes it impossible for the Church to enter into dialogue with other Christians, but perhaps more importantly with other religions. The principle of charity encourages us to take seriously 'dialogue', 'communication' and 'consensus', and hence one can summarise the dialogue between Rorty and Vattimo in terms of the search for solidarity, charity and irony (Zabala, 2005: 13). The Church is useful insofar as it promotes values such as charity that secular pragmatists can also embrace. In my terms, this dialogue represents the modern concern to make the mysterious intelligible and to convert the ineffable into the effable. Such a 'theological turn' is very much in line with the spirit of a democratic age.

Rorty's notion of 'religion' is problematic because it consists primarily of belief. He appears to have no understanding of religion as a way of life involving emotions, practices, collective actions, rituals, material objects and sacred persons (saints, holy men, witch doctors and so forth). His view about religion is that much of the theology of traditional religions is no longer plausible in the modern world, but are people committed to the religious life simply on the basis of plausible beliefs? We can contrast Rorty with Ludwig Wittgenstein who opposed any attempt to treat religion as primarily about (erroneous) beliefs and had a greater understanding of religion as a way of life, but a form of life, the beliefs of which are ultimately outside normal communication (Wittgenstein, 1979). Rorty can make religion and pragmatic philosophy compatible only by robbing the former of its distinctive contents (Smith, 2005). In my terms, Rorty can appreciate the importance of religion in human society only be expunging the numinous experiences of the holy or the *mysterium tremendum* and there is little room in his account of religion for the idea that one might be converted as a result of some overwhelming emotional and transformative experience. There is no role here for the miraculous or transcendent since Rorty's liberal humanist would be more open to intellectual persuasion than to an experience of the Sublime. For Durkheim by contrast, it was the collective force of social life that gives mythology and cosmology their authority as a result of collective effervescence. In many modern societies, it is when religion becomes deeply bound up with, for example, movements of national resistance – as in the Solidarity movement in Poland – that these collective emotions have overriding power and relevance. In summary, it is difficult to see how religion can survive without some notion or experience of the ineffable nature of the sacred, but it is exactly this aspect of religion that Rorty's version of pragmatism rules out, because ineffability does not sit comfortably within his secular democratic commitments.

From the ineffable to the effable

It is perhaps appropriate here to step back in order to contemplate in more depth what the contrast between the ineffable and the effable really entails. The word 'effable' (*effabilis*) is from *effari* or to speak out. In English usage, effable survived until the mid-seventeenth century, but its presence is now archaic. It exists only as the hypothetical counterpart to the ineffable, but I want to suggest that this absent term is central to modern culture. The ineffable refers to that which cannot be uttered or communicated; it is the unspeakable, unutterable and inexpressible. In its jurisprudential meaning, it indicates that which can be lawfully put into words. In religious language, faith is an essential aspect of religious life, since the actions of God are ultimately unknowable by reason alone. In the Jewish tradition, even the name of God is secret; hence the use of the Tetragrammaton or YHVH or Yahweh. The worship of Yahweh was 'aniconic' because it is without images or icons and His personal identity could never be known. In Islam, this tradition is perhaps taken even further, where *al-Ilah* is literally 'The God' – the eternal and uncreated Creator of the universe. Allah however is not a personal name.

Because the communication of sacred reality is typically ineffable, religious traditions require a stratum of intermediaries (such as theologians and other intellectuals) to interpret and translate the ineffable meaning of sacred reality. Over time these divine messages are encoded into languages that we now regard as dead and hence popular religious intellectuals, often employing modern media, become the intermediaries (talk-show hosts, opinion leaders, journalists, TV personalities and the like) who make the ineffable effable.

Literacy is a key issue in matters of religious authority. As a result, translations of the Divine Word were often unavailable to the uneducated masses who depend either on a literary elite that exerts its hegemonic control over divine speech or on popular teachers or agents (such as spirit mediums) to render the invisible visible. Often there was a physical barrier between the sacerdotal priesthood and the laity as represented by the rood screen in mediaeval churches. With the erection of architecturally imposing and splendid rood screens, by the fourteenth century the ineffable had also become the largely inaudible. The ineffable nature of the sacred word in these traditional societies was the intellectual property of elites who could read and interpret the sacred language (Hebrew, Latin, Arabic or Sanskrit). In Muslim societies, these literate elites stood in opposition to popular movements such as Sufism in which the ineffable was rendered intelligible through ecstatic experiences, dance, trance or divination. Popular religion has through the ages involved some attempt to access the ineffable speech of the sacred realm often through material objects – such as amulets and relics – and through the services of popular religious practitioners who were themselves often illiterate.

Birds have enjoyed the capacity of translating the ineffable. In the New Testament Gospel according to Luke 3 (23), 'the Holy Ghost descended in bodily shape like a dove upon him, and a voice came from heaven, which said "Thou art my beloved Son; in thee I am well pleased"'. This baptismal event was crystallised in the idea of the Holy Spirit as the Paraclete (*parakletos*), the advocate of the justified soul against evil forces, and hence the Holy Ghost is the Comforter of the Church. The Spirit or *pneuma* that entered Jesus in the form of a dove gave him the authority to throw out demons. The symbolism of birds – dove and eagle – remained important in early Christianity, often as intermediaries, but such symbolism from a pastoral community makes relatively little sense in the modern world where 'the cosmic liturgy, the mystery of nature's participation in the Christological drama, have become inaccessible to Christians living in a modern city' (Eliade, 1959: 179). The basic iconography of Christian soteriology – harvests, shepherds, the fishermen, the lamb and the dove – have lost their metaphorical force in a post-industrial system, because the shared experience of a pastoral or agricultural society no longer exists (Turner, 2001, 2009c). This rather obvious fact has been the driving force behind all theological attempts to demythologise Christianity.

One illustration of this transformation may be seen in the role of angels in the history of the Church. Angels were central in the early Church as mediums that brought messages to human beings from a different reality. They are not human, they appear at some point of crisis, and they typically descend to bring their message from on high. These somewhat austere and powerful beings do not fit easily into a less hierarchical culture of electronic communication, but the disappearance of angels as messengers can in fact be traced to the twelfth century when historians detect a shift away from the harsh world of Norman feudalism. As the angels withered away they were replaced by the Virgin Mary, and in the new legends of the Virgin we have entered a world in which the individual seeks comfort from a supernatural person whose companionship is tender and pure (Brown, 1975: 146). Unlike the mediaeval saints, the Virgin left no relics, but offered only her milk as spiritual nourishment. If angels survive at all in modern society, it is only in the shape of a 'guardian angel' or as the Madonna of popular music. Outside the context of American popular culture, the Virgin survives as a very powerful political symbol. The Virgin Mary is crucial in communicating between the wretched of this Earth (especially women and children) and a merciful God as we find in countries such as Mexico and the Philippines in the Virgin of Guadalupe, or the Black Madonna in Poland (Warner, 1976).

In the modern period there is a higher level of literacy and consequently a greater degree of the democratisation of knowledge, and with access to knowledge through the Internet, the sharp distinction between the elite and the mass becomes blurred. In a democratic environment, the very idea that some truths are ineffable conflicts with the ethos of modern society in which everybody assumes a right to understand, to intelligibility or at least to have the relevant information. Democracy promotes plain speech, and political campaigns are based on personalities and slogans and not only on explicit policies. The control of ineffable knowledge is compromised and the whole idea of hierarchically organised wisdom evaporates. We are moving from an age of revelation to an age of information where everything is, at least in principle, effable. The resulting crisis of intellectual authority is perhaps the real meaning of secularisation and despite all the talk of 'resacralisation' the world of deep ineffability appears to be inescapably doomed.

In summary, we may argue that in the age of revelation, religious communication had certain key elements that defined what constituted 'religion'. First, the structure of the system of communication was essentially hierarchical in the literal sense that messages come down to Earth from above. However, the social structure behind these religious codes was also hierarchical, and there was typically an interaction between religious and political authority. Borrowing a term from Weber in Volume 1 of *Economy and Society* (1978: 54), we can describe this structure as a hierocracy. In Weber's terms, the state has a monopoly of force and the Church seeks to achieve a monopoly of grace. These two institutions need each other, because religion is mobilised to legitimise power, and religions need the patronage and protection of political institutions. Because religious communicative acts were necessarily tied to general structures of power, religious metaphors reflected the structures of power in a society. The metaphors of divinity tend to be couched in the language of an absolute monarchy in which divinity was referred to as 'My Lord' or *Adonay*.

The main change that has taken place with the growth of these information systems in the age of global media is that the power relationships between popular and virtuoso religion have been reversed. The historical struggle between popular and elite religion in the field of symbolic capital has given an important if unintended advantage to popular religious movements which can now bypass the hierarchal organisation of orthodox or official cultures. The new media, along with many other social changes, have brought about a democratisation of the systems of religious communication with respect to both codes and contents, thereby transforming the power and role of various intellectual elites.

Conclusion: globalisation, religion and subjectivity

The globalisation of religion may be said to take three forms (Cox, 2003). First, there are various global movements that are broadly revivalist, often retaining some notion of and commitment to institutionalised religion (whether it be a church, a mosque, a temple or a monastery) with an emphasis on orthodox beliefs that are imposed authoritatively. Within this revivalist tradition, there are conventional forms of fundamentalism, but there are also the Pentecostal and charismatic churches. Second, various forms of popular and folk religion continue. These are practised predominantly by the poorly educated, and they seek healing, comfort and riches from such traditional religious practices. Third, there is the emergence of new spiritualities that are heterodox, urban, commercialised forms of religiosity typically existing outside the conventional churches, and often appealing to the new middle classes in service sectors of the global economy. In particular the development of spirituality or what I shall call 'low-intensity religiosity' caters to the individual need for meaning, but these post-institutional forms of religion do not necessarily place high demands on the individual. Privatised forms of religious activity do not contribute significantly to the vitality of civil society, but simply provide subjective maintenance to the individual. The growth of consumer society has had a significant impact on religion in terms of providing models for the commodification of religious lifestyles, and much global religiosity involves complex combinations of spirituality, individualism and consumerism. There is no doubt that the locus of intellectual power has shifted, in the West at least, from ecclesiastical elites to secular intellectuals, but these intellectuals themselves are increasingly marginalised by the elites that control the media, the financial markets and consumer culture.

The principal characteristics of religion in modern society, especially Western society, are its individualism, the decline in the authority of traditional institutions (church and priesthood) and awareness that religious symbols are constructs (Bellah, 1964). Modernity appears to be wholly compatible with the growth of popular, deinstitutionalised, commercialised and largely post-Christian religion. In a differentiated global religious market, these segments of the religious market compete with each other and overlap. While the new spirituality is genuinely a consumerist religion, fundamentalism appears to challenge consumer (Western) values, However, even the most pious religious movements can become saturated with the new consumerism; for example, as the veil becomes a fashion object or when Christians promote supermarkets that behave according to Christian standards. Piety often involves selling a lifestyle based on special diets, alternative education, health regimes and religious tourism. Gender is a prominent feature of the new consumerist religiosity where women increasingly dominate the new spiritualities; women will be and to some extent already are the 'taste leaders' in the emergent global spiritual marketplace.

While globalisation theory has often overstated the importance of modern fundamentalism (as a critique of traditional and popular religiosity), perhaps the real effect of globalisation is the triumph of heterodox, commercial, hybrid popular religion over orthodox, authoritative professional versions of the spiritual life. Their ideological effects cannot be controlled by religious authorities, and they have a greater impact than official messages.

Throughout this chapter I have spoken of a parallel between the emotive individualism of the market and consumer sovereignty, the growth of emotivism in moral philosophy and the rise of spirituality. In this discussion 'parallel' has the same meaning as the chemistry metaphor in Weber's notion of 'elective affinity'. At the conclusion of this discussion, I can perhaps switch to the stronger notion of a causal narrative. The rise of literacy in the nineteenth century gave the majority of Western populations access to popular science and some understanding of the significance of secularism and Darwinism. These changes were adequately described in MacIntyre's *Secularization and Moral Change* (1967). From the 1870s until the aftermath of the First World War, there was a gradual decline in the authority of ecclesiastical institutions and their spokesmen. In the second half of the twentieth century, the rise of consumer society made the traditional theodicy of the Church – suffering, redemption and resurrection – increasingly irrelevant and this change was reflected in the sharp decline in recruitment to the priesthood. Religion has survived in the West in the form of spirituality which is a post-institutional, hybrid and individualistic religiosity. Spirituality is the religious parallel of the sovereign consumer. The paradox of a post-secular society is that religion is booming, while the sacred is in terminal decay (Luckmann, 1990).

References

Bell, D. (1974) *The Coming of Post-Industrial Society*, New York: Basic Books.

Bellah, R.N. (1964) 'Religious Evolution', *American Sociological Review* 29: 358–374.

Bourdieu, Pierre (1987) 'Legitimation and Structured Interest in Weber's Sociology of Religion', in Lash, S. and Whimster, S. (eds) *Max Weber, Rationality and Modernity*, London: Allen and Unwin, pp. 119–136.

Brown, P. (1975) 'Society and the Supernatural: A Medieval Change', *Daedalus* (spring): 133–151.

Bultmann, R. (1984) *New Testament and Mythology and Other Writings*, London: S.M. Ogden.

Casanova, J. (1994) *Public Religions in the Modern World*, Chicago, IL: University of Chicago Press.

Cox, H. (2003) 'Christianity', in Mark Juergensmeyer (ed.) *Global Religion: An Introduction*, Oxford: Oxford University Press, pp. 17–27.

Derrida, J. and Vattimo, G. (eds) (1998) *Religion*, Cambridge: Polity Press.

Durkheim, E. (2001) *The Elementary Forms of Religious Life*, Oxford: Oxford University Press.

Eliade, M. (1959) *The Sacred and the Profane. The Nature of Religion*, New York: Harper & Row.

Habermas, J. (2006) 'Religion in the Public Sphere', *European Journal of Philosophy* 14(1): 1–25.

Habermas, J. and Mendieta, E. (2002) *Religion and Rationality. Essays on Reason, God, and Modernity*, Cambridge, MIT: MIT Press.

Gellner, E. (1988) *Plough, Sword and Book. The Structure of Human History*, London: Paladin.

Kant, I. (1998) *Religion within the Boundaries of Mere Reason*, Cambridge: Cambridge University Press.

Lazreg, M. (2009) *Questioning the Veil. Open Letters to Muslim Women*, Princeton, NJ: Princeton University Press.

Luckmann, T. (1990) 'Shrinking Transcendence, Expanding Religion?', *Sociological Analysis* 5: 127–138.

MacIntyre, A. (1967) *Secularization and Moral Change*, Oxford: Oxford University Press.
—— (1984) *After Virtue*, Notre Dame, IN: University of Notre Dame Press.
MacIntyre, A. and Ricoeur, P. (1969) *The Religious Significance of Atheism*, New York and London: Columbia University Press.
Macken, J. (1990) *The Autonomy Theme in the Church Dogmatics*, Cambridge: Cambridge University Press.
Mackintosh, H.R. (1937) *Types of Modern Theology. Schleiermacher to Barth*, London: James Nisbet.
Mahmood, S. (2005) *Politics of Piety. The Islamic Revival and the Feminist Subject*, Princeton, NJ: Princeton University Press.
Mandaville, P. (2001) *Transnational Muslim Politics. Reimagining the Umma*, London and New York: Routledge.
Martin, D. (2002) *Pentecostalism: The World their Parish*, Oxford: Blackwell.
McCormack, B.L. (1995) *Karl Barth's Critically Realistic Dialectical Theology. Its Genesis and Development 1909–1936*, Oxford: Clarendon Press.
Mernissi, F. (2003) *Beyond the Veil. Male–Female Dynamics in Muslim Society*, London: Saqi Books.
Nietzsche, F. (1968) *The Will to Power*, New York: Oxford University Press.
Otto, R. (1959) *The Idea of the Holy*, Harmondsworth: Penguin.
Rey, T. (2007) *Bourdieu on Religion: Imposing Faith and Legitimacy*, London: Equinox Publishing.
Robinson, J. (1963) *Honest to God*, Westminster: John Knox Press.
Roof, W.C. (1993) *A Generation of Seekers: The Spiritual Journeys of the Baby Boom Generation*, San Francisco, CA: Harper.
—— (1999) *Spiritual Marketplace. Baby Boomers and the Remaking of American Religion*, Princeton, NJ, and Oxford: Princeton University Press.
Rorty, R. (2005) 'Anticlericalism and Atheism', in Zabala, S. (ed.) *The Future of Religion*, New York: Columbia University Press, pp. 29–41.
Schleiermacher, F. (1996) *On Religion: Speeches to its Cultured Despisers*, Cambridge: Cambridge University Press.
Schochet, G.J. (1975) *Patriarchalism in Political Thought*, Oxford: Oxford University Press.
Smith, G. (2008) *A Short History of Secularism*, London: I.B. Taurus.
Smith, N.H. (2005) 'Rorty on Religion and Hope', *Inquiry* 48(1): 76–98.
Taylor, C. (2002) *Varieties of Religion Today*, Cambridge, MA: Harvard University Press.
—— (2007) *The Secular Age*, Cambridge, MA: The Belknap Press of Harvard University Press.
Tillich, P. (1956) *The New Being*, London: SCM Press.
Tong, J.K.-C. and Turner, B.S. (2008) 'Women, Piety and Practice: A Study of Women and Religious Practice in Malaysia', *Contemporary Islam* 2(1): 41–59.
Turner, B.S. (2001) 'The End(s) of Humanity: Vulnerability and the Metaphors of Membership', *The Hedgehog Review* 3(2): 7–32.
—— (2009a) *Can we Live Forever? A Sociological and Moral Inquiry*, London: Anthem Press.
—— (2009b) 'Goods not Gods: new spiritualities, consumerism and Religious Markets', in Rees Jones, I., Higgs, P. and Erkerdt, D.J. (eds) *Consumption and Generational Change. The Rise of Consumer Lifestyles*, New Brunswick, NJ: Transaction Publishers, pp. 37–62.

—— (2009c) 'Religious Speech. The Ineffable Nature of Religious Communication in the Information Age', *Theory Culture and Society* 25 (7–8): 219–235.
Vattimo, G. (2005) 'The Age of Interpretation', in Zabala, S. (ed.) *The Future of Religion*, New York: Columbia University Press, pp. 42–54.
Volpi, F. and Turner, B.S. (2007) 'Making Islamic Authority Matter', *Theory Culture and Society* 24 (2): 1–19.
Warner, M. (1976) *Alone of All Her Sex. The Myth and the Cult of the Virgin Mary*, New York: Alfred A. Knopf.
Weber, M. (1966) *Sociology of Religion*, London: Methuen.
—— (1978) *Economy and Society. Outline of Interpretative Sociology*, Berkeley: University of California Press, two vols.
—— (2002) *The Protestant Ethic and the Spirit of Capitalism*, Harmondsworth: Penguin.
Wittgenstein, L. (1979) 'Remarks on Frazer's *Golden Bough*', in *Wittgenstein: Sources and Perspectives*, edited by C.G. Luckhardt, Ithaca, NY: Cornell University Press.
Zabala, S. (ed.) (2005) *The Future of Religion. Richard Rorty and Gianni Vattimo*, New York: Columbia University Press.
Zubrzycki, G. (2006) *The Crosses of Auschwitz. Nationalism and Religion in Post-Communist Poland*, Chicago, IL, and London: University of Chicago Press.

7 Social fluidity

The politics of a theoretical model

Fernando J. García Selgas

Introduction

From the early 1980s (Baudrillard, Bell, Berman) to the current debates about social fluidity (Castells, Bauman), the specific features of our societies have been increasingly perceived as fragmentation, time–space stretching and condensation, dedifferentiation, mobility, and so on. This way of looking at society has been summarized in the idea that society is now fragile and unstable, but not necessarily soft; it is like a fluid reality.

This idea became more appealing with the confluence of historical events and scientific transformations. Historical processes like the globalization of capital flow, a more flexible workforce, technological revolutions, the hegemonic presence of media culture, and new social movements have acted as "social-reality-mixers". At the same time, a growing number of conceptual or theoretical changes have enabled us to dissolve basic, traditional, unquestioned dichotomies (action/structure, global/local, micro/macro, essential/constructed, etc.) and shift from movement to fluidity as historically regulating ideas.[1] Last but not least, the work by the aforementioned authors as well as Harvey and Jameson may be seen to converge in the claim[2] that we are immersed in a historical process of radical fluidification which is affecting the very foundations of our societies, i.e., it is affecting their:

- Leading dynamics: instead of being triggered by a constant alternation between breaking away from the past and building the future, between the solidity of established institutions and the evanescence of a utopia, we are now immersed in a complex mixture of different, alternative presents where social development is more a mixing process than an innovative practice.
- Protagonists: nation-states are opened and weakened by transnational and local forces alike; there is no longer any fixed value for capital, the supposed subject of capitalist development; working class and individuals, the heroes of Marxist and liberal narratives, seem to be drowning in the wishing well of consumerism.

- Basic materials: when the basic social stuff combines flows of capital, information, people, commodities, etc., not only the main characters but also social action, identities and social time–space all become unstable and fluid.

As in previous periods in the history of social sciences when new phenomena and concepts emerged together,[3] two apparently different sets of problems merged.

On the one hand, there were questions about the depth and radicalism of these social changes: Are they bringing out a new social reality, a new social ontology? Are we confronting a new way of becoming, being or deploying social structures, social agents, identities, social forces and dynamics? Do we have to think of social events in new terms? Do we need a new theoretical model? All of these issues were raised as questions of a scientific nature that need to be answered with scientific (i.e., empirical and conceptual) work.

On the other hand, it was not by mere coincidence that most social scientists involved in this research and debate about social fluidification had a Marxist bent in some way or another. Urgent questions were also emerging from this vision concerning political issues such as: Are we witnessing the decomposition of the working class and hence the end of any radical challenge to the core of the capitalist system? Is the fluidification of social forces dissolving any chance of social criticism or resistance? Are individualization and presentism untying social bonds? These queries were all posed as political questions requiring practical and strategic implementations more than simple words or observations, although some discursive movements may have been involved.

I am not going to deal at first hand with these two kinds of questions, but rather with the idea that they do not come apart. I shall argue that the first questions cannot be answered without dealing with the second ones; that there is no social or sociological theory without political implications. This chapter strives to illustrate the politics of a sociological theory by focusing on how a specific theoretical model of social reality (as fluid reality) enacts specific political questions, ideas and practices. To do so, we first need a short outline of the theory and its relations with other competing theories in the field of social sciences (see section 1). I will then show how politics, primarily the politics of dissent, are internally linked to this theory (see section 2).

It is true that after more than four decades of work on the new social studies of techno-science and the post-Kuhnian philosophy of science, political implications are commonly assumed in the disciplinarian, methodological, technological and applied levels of sciences, primarily in the case of the social sciences. However, this is not the kind of agreement that we find about the political implications of the theoretical level in sciences. Not much has been said about how the development of concepts and the production of theoretical models, which underlie every hypothesis, statement and research technology, mingle with political issues.[4] This is therefore the main concern of this chapter. I will argue that theoretical work in the social sciences articulates political possibilities, issues and actions. I will describe the politics of a theoretical model of social reality as a fluid reality, but in order to do so, without being driven away from the theoretical level of knowledge, I shall take the meta-theoretical research path instead of an empirical or historical approach.

1 The theoretical model of social fluidity

The conceptual movements and historical analyses mentioned at the beginning of this chapter gave birth to a fuzzy web of concepts and ideas about social reality as fluid reality. It was a kind of rudimentary theory claiming that now social existence is increasingly fluid, that nowadays social reality is a complex and fluid assemblage of heterogeneous ingredients more than a material substance or a formal structure. A new theoretical model of social reality (i.e., a new overview of the kind of stuff we are dealing with in social sciences, a new outline of the ontology of social reality) started to emerge,[5] confronting the two traditional competing social science models. A brief review of these two models and a schematic presentation of the new one should suffice to introduce the case at hand.

Two hegemonic theoretical models have prevailed since the first general claims about social reality. On the one hand, there is a traditional vision of society as an aggregate of independent, primarily human elements or substances such as individuals or national states. This is a substantialist and somehow materialist vision related to the idea of a communitarian or political nature of human beings (human sociability). The social realm is regarded as the relations among subjects (mostly individuals) that usually take place in the space of a political community such as the nation-state. This vision may be found in the Aristotelian definition of humans as "political animals", in Weberian methodological individualism and in rational choice theories. On the other hand, modernity was accompanied by the emergence of a second model in which society was thought of as a structure or system of determinations or communications that goes beyond individuals but usually refers to the human community. This is a formalist view in which society is seen as a *form* (structure) in which individual or collective agents find themselves in specific positions, oppositions and differences that set conditions and limits to their lives. We can find this vision in historical materialism (economic capital seen as the determining structure), neo-functionalists and the work by Luhmann (functional or autopoietic systems).

Both models assume that there is some kind of basic social solidity underlying social changes and mutations granted by either a systemic form or by the nature, essence or law of development of the agents involved. In contrast, our model finds fluidity instead of any basic solidity; instead of any formal or substantial constitution of social reality, it finds a material relationality in which every basic ingredient of social reality (every agent, for example) is constituted and social action is opened up to non-human beings.

Each model has a different degree of accuracy and plausibility in the description of our current social reality. I would say that although we cannot speak of this reality without conceptual or theoretical assumptions, insofar as we have undergone a deep process of fluidification, the third model makes the best match. Our main concern, however, is to see the political implications that inhabit the new theoretical model.[6] To reach this goal, we first need a draft of the specific focal lines of this theoretical model and their perspective on our social world as a fluid reality:

1 This model speaks of social reality as *unstable multiplicities*, instead of unitary or solid substances or structures. It is claimed that most social forms and formations, from home and work to politics, are unstable, contingent and fast-moving. "Shaping them is easier than keeping them in shape" says Bauman (2000: 8). It does not mean that they are soft, easy or weak. Big cities are nowadays good examples: they have been filled with no-place spaces (airports, malls) and other spaces where people cross paths without meeting. In these cities, social bonds become unstable and fluid. At the same time, to become a (collective or individual) social agent in this kind of social reality, we need the assemblage of multiple, changing, heterogeneous ingredients, from bodies and technologies to systems of signs and emotional syntax. Social facts and agents are not out there waiting to be discovered or described; they are part and consequence of some heterogeneous assemblage (including scientific discourses); they are, in this sense, unstable multiplicities.
2 The grounding of (fluid) social entities is not located in any kind of nature or determinant form but in *mutual relationality*. Social existences are therefore described as being mostly built on an unstable, ongoing relationship among multiple, heterogeneous ingredients: they are nurtured by relationality. However, unlike the other models, nothing in this relationality is pre-formed and there are no primary elements such as intentions, capital accumulation or system/environment distinctions. All the assembled ingredients, including networks, individuals, regularities and materialities, are mutually shaped in that relationship. The mutual, relational and recursive constitution of a set of different ingredients (linguistic, technological, material, imaginary, flesh and blood, etc.) is required for the emergence of a social agent or institution, including a large one like a nation-state. Accordingly, social borders remain wide open and all social entities remain under construction. Their challenge is not "to be or not to be", but "how to become", "ways of becoming".

3 Instead of substantialist isolation or systemic differentiation, the model of social fluidity brings in a *sociality open to non-human beings.* Current social realities are therefore regarded as a sort of complex, open mixture where, for example, technology becomes "society made durable" (Latour, 1991) and politics turns into social antagonism and the art of the possible (Mouffe, 1997). Promiscuity thus replaces unity as the constructive logic. This does not mean, however, that society becomes everything, because at the same time, its ingredients are seen as being inside and outside the social realm, they have partial autonomy, with specific logics (logic of efficiency in technology and logic of hegemony in politics). It means that, contrary to one of the old sociological beliefs, human beings, actions, values and relationships are not always at the centre of social realities. These realities also appear to be constituted by, in and through objects and other non-human beings, their effects and relationships.

4 The social fluidity model does not entail any ontological closure, be it substantial or formal, material or conceptual, but rather some kind of *porous borders.* This porosity emerges in the internal connections between different kinds of existence (symbolic, material, emotional, etc.), but also in the oxymoronic crossing of fluid and solid logics.

Consequently, with this model, social solidity may sometimes be seen to hide an unstable, incomplete, precarious and changing arrangement of heterogeneous ingredients that turns out to be fluid. A wonderful case study by De Laet and Mol (2000) on a water pump used all over Zimbabwe shows how a solid machine like a water pump is fluid in three different ways: it has several identities, as a mechanical object, a hydraulic system, a health agent, a community resource and even as a national identity, with different but interconnected and unstable limits; in each of these identities there is no sharp distinction from the surroundings, no closed borders; and the evaluation of any of these identities is not a matter of yes or no, because it depends on changing criteria.

At other times, this model help us to see social solidity emerging from the most fluid social processes and events, as in the case of gender violence: in the most tragic events of gender violence, we can see a sort of "macho obduracy" emerging from the current, radical fluidification of gender identities and relationships.

5 In line with the other two theoretical models, the fluidity model sees social forms and social order as ontological conditions of social existence, not mobility and speed. As in the case of the study of fluid realities in natural sciences, to speak of unstable shapes does not preclude speaking of some sort of stable form. Here it is even easier because a *fluid social form* is seen not as a network, a system, or any other purely formal presence, but as *material, contingent, open and disputed articulation* of relationships, fuzzy borders and porous limits, with complex

but available stabilizations. In this way, we can consider how most nation-states are now able to retain an efficient and powerful formation through a challenged ability to articulate transnational agencies, local interests and transversal coalitions of economic, social or cultural capital, and so on. Besides their materiality, fluid social forms are not described as given but as part and consequence of what may be called *ordering* or *stabilizing processes*. Of course, we can retain the idea that these processes have something to do with patterns, regularities, norms, institutions, or even a sort of solidity. However, they will always be deployed in the mi(d)st of a complex, fluid relationship of different works, multiple entities and considerable consumption of various energies: ordering and fluidification go hand in hand. Let us think, for example, of how a social institution like marriage may be performatively stabilized by the same practices that challenge it, as in the case of gay marriages.

This brief presentation of the core of the theoretical model of social fluidity may be summarized by saying that it speaks of current social realities, in plural, as different, complex and unstable relationships among multiple, heterogeneous, and mutually constitutive ingredients.

2 The politics of a theoretical model

Our next goal is the political implications and attachments of this theoretical model. We will see how specific political insights and practices arise with the new conceptual model, and vice versa. It should, incidentally, be a good example of how theoretical questions about the nature of social reality go hand in hand with political issues and practices.

Looking at the historical changes mentioned initially from the perspective of leading social scientists such as Sassen and Beck, it is easier to arrive at a common and simplified picture of current social reality as unbound social bonds, individualisms, unstable regulations, flexible and global capitalism, hard and soft powers, and sporadic or irrelevant political actions against these powers. Perhaps it is not a bad description to start with, but, if we bear in mind the theoretical model of social fluidity, we can not only understand this picture but also detect and eventually reinforce different spaces for political criticism and resistance that are unnoticed in other perspectives.

Other scientists studying the same process of social fluidification have located social fluidity only at the more general or structural level, in relation to big issues such as new information technologies or global capitalism, while individuals or even nations are seen as just trying to cope with it.[7] As a result, it is presented as a world of small, inflexible agents (be it individuals or nations) struggling against big, flexible, fast-moving powers, where politics is reduced to no more than attitudinal or dispositional movements (politics of identity) and the conquest and exercise of power (governance and election campaigns). However, if what we have in mind is the model of a fluid social reality (i.e., an unstable, open, contested reality), we should be able to get a more complex picture in which social reality is open almost everywhere to antagonism and resistance, the micro-macro distance is part and consequence of complex, flexible, ongoing practices, and politics may be not just a matter of taking sides but also a question involving potential, antagonistic and semiotic-material forces. Ultimately, it will give us new political understanding and possibilities.

2.1 The political dimensions exposed by the model of social fluidity

We can take a first step in this direction by looking inside the basic ingredients of current social reality from the fluid model perspective, which allows us to see political spaces, open to criticism and dissent, that have been overshadowed by other socio-political understandings. By checking the usual basic elements of society (social actors, situations and structures) from the theoretical model of social fluidity (i.e., checking (1) agents, (2) social time–spaces, and (3) their respective stabilizations), we will see how politics merge in every one of them.

1 Due to increasing fragmentation and a wide variety of problems that are closely related to the mixture and conflictive instability of current collective identities, it is now troublesome to live with and understand personal identities. Gender (mainly masculine) identities in Western countries are a good example of this situation. How can we manage, or even name, the flux of social positions, learned dispositions and personal identifications and decisions that form our subjectivities yet remain in a kind of contradiction with what we were supposed to be and expect as capable socialized agents? Today, personal identities are no longer a warm refuge or a frame of reference but a practical problem (How can we build or stabilize them? What possibilities do we have for them? What can we do with them?), and a conceptual question (What kind of thing are they?).

   These problems are worsened due to general circumstances in the current shaping process of identities or subject positions. For example, our consumer life masks our experience with illusions of choice and replacement, seriously hindering their sedimentation and our possibilities of taking root, and generating a deep feeling of isolation. Oddly as it may seem, this is worse among individuals with less consumer capacity.

   As a consequence of these and other similar processes such as the dissolution of the social bonds that have been fuelling emancipation practices (e.g., class consciousness), it seems that social agents (individuals, but also states) are losing what has usually been considered as political agency, that there is a lack of (traditional) politics. However, when we consider these processes using the fluidity model, social agencies, identities and subject positions appear themselves to be unstable processes open to antagonisms and contestation, revealing their political nature, a view that was previously announced by the feminist slogan: "Personal is political."

2 A similarly disputed and, in this sense, political character may now be perceived in the second basic ingredient, i.e., in the case of hegemonic social time–space.

Research by many social scientists into time–space stretching and compression, from Giddens to Harvey and Castells, has proved that we are in a time–space of flux, where local–global connections and the relations between past, present and future are increasingly complex. We can speak of a fluid topology (Mol and Law, 1994) where everything is interconnected and de-differentiation is the marching tune, but we can still detect some important differences.

For example, migrants with the right to vote in two countries, internal and external (regional) dissent with the state, the increasing power of supranational institutions, and other (fluid) tendencies, are blurring the national state space as *the* political territory, erasing what has been traditionally regarded as the more specific political arena. In addition, the global economy is not only increasing fragmentation, insecurity and inequality in these spaces, but is also eliminating any external vanishing point in time–spaces that could act as a reference point (or utopia) for an alternative to the hegemonic order. As a result, it becomes extremely difficult to find a classical political space for the deployment of critical practices or discourses.

However, looking from the fluidity perspective, we will notice that the same institutions which blur the outline of traditional politics spaces are unstable, suffer from internal dissent and create new political arenas. For example, the unstable, mobile and transnational time–space produced by most migrants seems to be configuring a more complex, fluid relationship with the state(s), in which the very condition of citizenship is not just an assumption or a goal, but a disputed object with its limits, implications and conditions open to debate and struggle. We can thus perceive a new, different political time–space where local and personal interests criss-cross with global finances, new communication technologies, diverse legislations, and so on.

3 Social stabilization or ordering processes, the third basic ingredient, are also perceived as disputed or political when viewed through the lens of the fluidity model. We just have to recall how this model claims that (fluid) social stabilizations are material, contingent and disputed articulations of relationships, fuzzy borders and porous limits: they are complex but available. From this perspective, we can recognize at least three different sources of disputed or political issues in social (fluid) stabilizations.

According to the fluidity model, the actual performance of any social action has the potential of making a difference at any given moment-position in a (fluid) stabilizing process, not only in a semiotic sense but in a material one as well. Consequently, there are not innocent social actions, including the case of the more subordinate agents: it is the end of innocence. At the same time, it is also the end of blind obedience, including civil servants, the military and the clergy, and the beginning of widespread but highly complex accountability. We arrive at a similar vision from the claim, made by this model, that institutions are contingent and disputed assemblages of open relationships, because in this case nobody knows beforehand whether or not a particular operation (by an NGO, for example) will ultimately be unfair, cruel, etc. It will depend on the oncoming, disputed and impossible-to-monitor connections of that operation.

Beyond the uneven, contested distributions of all kinds of material and social resources produced by the stabilizing processes, the fluidity perspective highlights the complex and controversial confluence of open (fluid), different stabilization or structuring processes. This is particularly visible in the case of the many disputes over representations that flourish everywhere, from the most mundane to the most sublime controversies, from "How are we classified?" to "Who speaks for whom?" or the performativity of all representation. Sometimes we can even speak of "the wars of representation" that take place everywhere, from science to parliament and biopolitics (see Haraway, 1992). Cultural, semiotic and knowledge fields thus become political arenas in the fluidity model. A growing number of works of art displayed at the most prestigious international exhibitions are a good example, as they become involved in the conflicts of representation by denouncing new forms of oppression or advocating specific groups, rights or political stances. That is the case of Ines Doujak's work in 2007, Kassell Documenta, in which idyllic gardens and ornamental decorations are filled with information about the "global biopiracy" of chemical transnational corporations and explicit images of drag kings and queens, confronting us with one of the latest forms of colonialism and defending different, unashamed sexual options. She manages to bring together both issues and make a strong general claim for biodiversity in life and culture, a clear example of (bio) political action in the fluid wars of representation, where multiple (and fluid) stabilization processes meet.[8]

Finally, the fluid model helps us to realize that we are looking at a thin layer of ice covering an unstable, ongoing, complex social flux when it claims that although we usually regard social institutions as the established or fixed scenario, they are the ongoing result of complex, unstable relationships between multiple, heterogeneous and mutually constitutive ingredients. From the perspective of fluidity model, social institutions appear as regularized processes built on a large amount of

> different, interconnected works within which struggles of interests, the exercise of power and antagonism reigns. As Law has said (1994: 95), the most solid, large and powerful social institutions are such because they (primarily those who profit from them) strive to erase the work of others (subordinates), silence underlying disputes, and freeze out the implicated social networks.[9] We can also see this internal political nature of social fluidity stabilizations by simply looking at many of our current, more quotidian social institutions, be it at the local (multiple and open forms of Western families), national (questioning of merit-based school systems) or global level (frequent disobedience of international law). They are political, basically because the multiplicities of practices and ingredients that make them possible translate them into unstable or disputed social forms at the same time.

The above arguments may be summarized by saying that from the fluidity perspective, most ingredients of current social reality are entangled with disputes, oppositions, antagonisms, interest, and other political issues and problems. However, this picture could also be seen as the disappearance of potential criticism or resistance. The question is then to show how those potentialities, and the agencies that can actualize them, become visible and possible when considered from the fluidity perspective (see section 2.2), and how all of this is connected to the renewal of our political discourse (see section 2.3).

2.2 When the fluidity perspective makes political agency and criticism visible

We agree with Bauman when he states (2000: 85–88) that the fluidification of identities and the instability of its references facilitates the fact that nowadays, the control and the power exercised over us does not follow the panoptic (vigilant), disciplinary form but rather the seductive mode of the "synoptic": we are moulded to watch the show, be entertained by it and live as mere spectators.[10] Instead of a few monitoring the majority, now it is the majority that watches, heeds and constantly follows the few, who, instead of being chosen, are contingent and temporally coopted. It makes no difference whether these few are artists, sportspeople or politicians; they all fall into the category of celebrities. It is a control system which fits perfectly within a social order that is mediated and constituted by mass consumption (especially service and cultural commodities) and media-driven information and entertainment.[11] The overwhelming synoptic presence of offers, shows and proposals, impossible to unhook from, seems to cause a paralysis in our critical capacity, an emptying of our will, our self, and ultimately, the agency capable of resisting. Let us check these two likely consequences (lack of criticism (1) and political agency (2)).

1 The paralysis of critical practice shows up in a synoptic or control system of domination only if we continue to regard "social or political criticism" in reference to the concept of revolution. Revolutions require the possibility of radical transformation of a stable, closed and oppressive order, or at least of storming the bastions of power,[12] but, at least from the fluidity perspective, it no longer seems possible to locate the control tower from where we are watched, or the castle or boardroom where the events are planned.

From this perspective, what is coming to an end is the modern or enlightened mode of criticism or resistance deployment, i.e., the mode of the destruction/construction dynamics, the revolutionary dynamics, the modern imagery of appearing to be *ex novo*, breaking off from traditions. On the one hand, according to the fluidity model, this is not the only mode of criticism or political resistance, and now it is not the main one either, because fluidification has affected not only the social order but also social disorder and social criticism. Besides, from this perspective we can realize that the general impossibility of happy endings (the after-revolution) or utopian guides does not preclude alternatives that support criticism of the existing order. On the other hand, unlike other theoretical models of a structuralist bent that also speak of some sort of fluid criticism,[13] the fluidity model claims that social order is fluid (unstable) as well. Looking through the conceptual lens of this theoretical model, we can detect on the one hand conservative or control effects in the practice of criticism or resistance, and on the other hand, current dissidence, criticism and resistance within the dominant dynamics themselves, because both of them are once again considered unstable and disputed, i.e., the former are not innocent and the latter are not free from criticism and opposition.

A good example of these fluid (complex) relationships between control and resistance may be found in the case of the current predominance of sign value, the predominance of image over merchandise and its monetary values,[14] a core element in the fluidification of social reality and the dismantling of some of its critical instances. It is precisely the malleability, the ephemerality and the power of images that enables them to also become means of struggle and resistance, as we can see in urban graffiti and hip-hop in low-income neighbourhoods or in the use of the black Zapata balaclava. However, they are susceptible to subsequent marketing and reabsorption by hegemonic seduction/domination mechanisms. Fluidity also affects criticism, generating its instability and complexity, but not its inexistence or impossibility.

2 The second likely consequence of the control system has been enunciated by Bauman (2000: 132–133) when he states that the most important, most difficult question today is not "What to do?", as it was at the start of the last century (Lenin), but "Who is going to do it?" In fact, this question is an inevitable political burden in the analyses of social agency

as a fluid. Remember that in the fluidity model, there are no essential elements or units of social reality, unlike individuals, states and means of production in the substantialist model. Everything is under construction, open to alternatives and controversies and, in this sense, political.

It is no surprise that in the current historical situation,[15] doubts about political agency, particularly progressive agency, arise because the most relevant political agents of modern societies such as the individual (the entrepreneur, the hero, the free thinker and other liberal figures), the proletarian social class and the state are fully occupied just trying to survive longer and not to become dissolved themselves. Even the kinds of political activities they have been practising (primarily the governance of national territory and economy) are increasingly overflowed by transnational fluxes and the growing importance of "life-politics" (Giddens).

However, to some extent the problem is in the eyes of the beholder. If, despite a profound transformation in social existence (fluidification), we keep grasping at the same old images about socio-political agency, such as actions in the public arena planned by individual or collective consciousness, we will be looking for the same kinds of political agencies, and, given that we will not find any such solid or autonomous social position, we will declare the end of politics.[16]

If, on the contrary, we follow the fluid perspective, we should be able to notice new or renewed forms of political agency. The multiple, unstable, changing forms of the anti-globalization movement is a well-known example of how to deploy political dissent and criticism of the liberal globalization trend; of how to intervene in very important political issues, even though the movement is no more than an open, fluid, and contradictory network of highly varied ingredients. A more clarifying case can may found in the complex struggles of interest over the law allowing homosexual marriages, passed by the Spanish Parliament on April 22, 2007. While not denying the conquest of some rights, this law could initially be regarded, as indeed some radical gay associations have done, as a step towards the integration of people living in alternative fringe lifestyles into the hegemonic social order.[17] In this sense, it could be a process which initially seemed to be a challenge to the hegemonic social order, but ended up as a way of expanding and reproducing that very same order. However, things are much more complicated, as we can see from the fluid perspective, and this very hegemonic order is in the process of fluidification as an unstable social order, in such a manner that the law is not only a means of integration but also a back door that is open to the profound questioning of our general concept of marriage. Under this interpretation, legislators have triggered (probably unintentionally) a deeply subversive movement against one of the basic ordering principles of our lives: the one built around the hegemonic ideas and forms of marriage and family.[18]

In both cases, the agents do not need either a clear awareness of what they are doing or a clear definition of who they are. For them and for everybody else, it is very hard to access this information within a traditional conceptual framework, because the complex and fluid nature of these political practices does not fit into it. Instead of being planned actions, they are a part and consequence of the assembly of different, changing ingredients, not all of which are human or symbolic.[19]

2.3 The politics of a political vocabulary

In the process of highlighting political agencies and criticisms from the fluidity perspective, as well as when we have seen the disputed and hence political facet of today's basic ingredients of (fluid) social reality, we have been questioning the established meanings of various political terms; we have being moving towards a refreshed political vocabulary. With the fluidity model – as well as with other theoretical models – there is a conceptual shift in our political vision, bringing in new views and possibilities. This renewal is a political stance as much as a semantic movement, as we can see in the fluidity model by recalling the external support for this renewal (1), as well as further development of these conceptual shifts (2).

1 External support comes from the nearest discourses. Sometimes it arrives in the form of an inspiring family of new political concepts such as the previously mentioned post-structuralists, although in this case we need to apply some discrimination.

 According to the fluidity model, all social and political reality is somehow fluid, unstable, and internally open to dissent. It therefore does not permit a dualist political perspective of solid political centers (like "Empire" in Negri and Hardt, 2004) exercising control over unstable and subjugated identities (their "Multitudes"), whose only chance is a fluid ("rhizomatic", in their terms), subversive kind of politics.[20] However, for that same reason it is possible to find space within the fluidity model for concepts of unstable, networked, and flexible political subject positions such as "multitude", which articulates different groups and interests, producing unity[21] without neglecting their differences. A new shared political imagination might be developed here which could eventually take part in the merger of political practices of resistance to different forms of subjugation, including the subjugation produced by any general form or strategy for these practices.

 At other times, external support comes through convergence around a new political notion, as in the case of the political complexity shown in the Spanish homosexual marriage law by the fluidity perspective, which pointed to the idea that radical political changes are nowadays more a question of "partial modifications" than ruptures or revolutions. Some of the ideas underlying this perspective such as the vision of social realities as multiple, never-finished entities,[22] the idea that power is pervasive and inhabits every social space and position, the denial of the pertinence of the (modern or enlightened) construction/destruction dynamics and discourses, or the perception of personal and collective identities as political battlefields, lead to the conclusion that "[I]t is in the accumulation of close partial modifications – that is a multiple, dissonant, multilateral and multidimensional collection of partial modifications – where the promise of a different future lies".[23] Accordingly, slow but deep swings in general sensibility, in common imaginary references, or in the confluence with non-human agency, for example, seem to open up a more fruitful path for social changes in a (fluid) social reality than an abrupt, reversing revolution (that turn upside down but mirror power relationships).[24]

It is around this and other related political concepts that we can see a supportive confluence of the fluidity perspective with a wide range of traditions, from post-Marxist political scientists to millenarian doctrines like Taoism.

When Mouffe (1997), for example, draws the political arena from the public sphere (*polis*) to the antagonisms, conflicts, and disputes (*polemos*) that pervade social life, she moves against the revolutionary (Enlightenment) concept of politics and in favor of a political agency formed by passionate mobilizations more than by rational strategies. This approach to the partial modification thesis was already visible in her germinal work with Laclau (1985), where, in trying to assimilate anti-essentialism, decentered subjects and other fluid realities in Marxism, they resume the notion of hegemony as a consequence of the contingency and pervasiveness of power, acknowledge the fluidity and conflictive constitution of all group identity, and promote an ideal radical democracy as an ongoing articulation of resistance to different forms of subjugation.

At the other extreme, we find Taoism as an early philosophy of fluid reality[25] and a guide for politicians in Eastern societies. It is not a philosophy of non-action (or quietness) but of allowing things to happen, of going with the flow. Among many other things, this means the following ideas, perfectly fitting into the political discourse of partial modifications: that non-human realities also belong to the political arena; that quite often, the only possible political action is not a hard confrontation but some soft means of change, like inverting or avoiding hegemonic flows; and that political action always affects the very agents involved in it.

2 Let us look now at the main trends or changes inscribed in the political vocabulary by the fluidity model. The first and most general trend is that all of these transformations in political concepts have been more a sort of re-signification or re-definition than a creation, because they always emerge within the above-mentioned focal lines of the

theoretical model of social fluidity, i.e., within the vision of heterogeneous, relational, multiple, open, hybrid social forms and realities. With respect to more specific changes, I can only present a blueprint of what is happening with three relevant concepts: "criticism", "emancipation", and "resistance", although I hope it will suffice to complete a rough outline of the political vision prompted by this theoretical model.

It is difficult to link the two foundational meanings of "*criticism*", "critique" (evaluation, examination or judgement) and "critical" (a difficult, defining, and unstable situation, a crisis), to a theoretical model in which there is no external point of reference to anchor any judgement, and "crisis" is the name of a constant and constituent condition of an unstable sociality. For this reason, the classic concept of "criticism", or better "objective criticism", a centerpiece in most spheres and paradigms of modern thought,[26] can no longer lie at the heart of the fluidity discourse. It is, however, a convenient ingredient in this discourse when it is understood as the practice of opposing or questioning some kinds of power relationships. Remember that for this perspective, every social action (including knowledge production) is located within some kind of power relationship. We may thus speak of (fluid) criticism as a way of highlighting the contingency of every social discourse and institution, and questioning its (self-) presentation as natural or necessary; as an internal resistance to current power/knowledge systems; as the unmasking of the semiotic chains of domination or violence that support subjugation structures, etc.

Modern political discourse established *emancipation* as the main goal and slogan of its purported progressive march, but it does not fit well with the fluidity model. At the core of emancipation, we can find the idea of complete autonomy for (all) human beings (i.e., an escape route from any absolute power, whether it be Mother Nature, (father) God, or (uncle) King); an escape that should be driven by human beings themselves and nothing else. It is as if we, or our systematic association, were a self-referent reality.[27] However, the fluidity model considers social reality as an unstable, heterogeneous complexity in which discourses, objects, technologies, other living beings, etc. are also a part and consequence, in addition to pervasive human beings and power relationships. This perspective does not depict human beings separately from other ingredients of social reality such as objects, animals, discourses, goddesses, and so on. From this perspective, we can no longer speak of emancipation as a movement towards human autonomy. We should rather speak of "*co-emancipation*" as a movement embracing all these social ingredients, moving towards common, mutual emancipation.[28]

Furthermore, when society is not reduced to civil society, politics is not reduced to the institutional or representational realm, and it is claimed that social fluxes which shape individuals are only possible because of the ongoing stabilization of additional non-human entities, the idea of a closed political space (parliament, nation, etc.), where a final agreement or reconciliation[29] among humans could be reached, seems to be wishful thinking and unreal. We arrive at similar ideas from the fact that (fluid) oppositional practices are seen to respond with the same arms to the aesthetic and technological practices performed to generate disaffection by synoptic, dominant powers, i.e., they enrol different sorts of mechanical, technological, semiotic and organic agents, beyond human beings.

"*Resistance*" was a marginal term in progressive political discourses of modernity until some of them turned into totalitarian ideologies. Little by little, it became synonymous with emancipation practices (Foucault) even when it was resistance to purported modernization (Giddens). Nowadays it remains in a state of constant resignification that makes any kind of definition pretentious. However, in the fluidity model, we find the exemplary cases of physical fluids, in which resistance is a very important and complex feature.[30] According to this reference, the fluidity model speaks of resistance as an important and complex feature of social reality that mixes opposition to structuring forces with sedimentation or stabilization processes; mixes so-called progressive and conservative politics; and combines oppressive and liberalizing effects. For that reason, (fluid) resistance has an oxymoronic emergence in subjugation process like consumerism (García Canclini) or the hegemonic, seductive, confusing media mix of information and entertainment (Kellner). It inhabits the fringes of social orderings as well as the very "belly of the beast". From this theoretical model, resistance may also be regarded as the ability of every (fluid) social agent to oppose any act of subjugation suffered by him/her/itself or alter (weakening, for example) any act of subjugation exercised through him/her/itself and over others. In this sense, resistance is a source of the politics of dissent and a sign of the end of innocence.

Concluding remarks

It is quite obvious that not all political discourses and terms linked to current social fluidification or even all scientific or academic discourses dealing with social fluidity are connected to criticism, co-emancipation, or resistance. In fact, some of them are hypnotic, fluid discourses, like the ones that may be found in new managerial handbooks or in financial engineering, striving to surf the craziest waves of capitalism and happily in tune with control strategies. However, what really matters now is that while focusing on the politics of dissent, we have clearly seen several internal and direct connections between the theoretical model of social fluidity and certain types of political discourses, possibilities, and practices.

When we have considered current forms of the basic ingredients of social reality (agents, situations, and social orderings) from the fluidity model, they seemed open to criticism, disputes, and dissent, even when other perspectives speak of the disappearance of political criticism or agency. The fluidity model prompts us to detect and eventually reinforce some sort of political involvement. It implements specific political forms and actions when making them visible and the object of discourse, which is not the same as constructing them.[31] These political forms do not include all kinds of existing politics though, but rather a specific web of political visions, terms, and practices: primarily they are politics understood as antagonism or dissent, as a contest of (im)possibilities, as partial modifications, and as a disputed renewal of political concepts and practices.

Other features or facets of this "fluid politics" may also be inferred from what I have said (e.g., the political character of discourses and ontologies in the fluidity model). As mentioned, the fluidity perspective states that social reality is multiple, unstable, and under construction; that it is a part and consequence of different forces, including scientific discourses.[32] The existence of multiple realities in the fluidity ontology manifests a constant game of dissent, interests, disputes, affinities, and compromises among different versions and (im)possibilities, which brings into play a range of political questions such as "Who should decide and how?" and "What kind of logic (technical, economic, etc.) governs the differences between options or versions here?".[33]

We can even speak of general features expressed in characteristics forms of fluid political agency and criticism. For example, because social entities are, according to this theoretical model, ruled by open relationships among changing, heterogeneous entities, their political entanglements turn towards "otherness" rather than towards equilibrium. A general trend that is echoed in some specific conditions of "fluid politics": the fluidification of the individual's boundaries (skin, home) generates actions and agents that are more interconnected and implicated with the environment and other voices; there is a growing, expanding resistance to any restriction to fluid existence; and the ever-changing relationship between order (stabilization) and disorder (fluidity) locates resistances and exercises of power in both.

However, our main conclusion is that the theoretical model of social fluidity enacts specific forms of political discourses and practices.[34] This claim allows us to, at least, point to a more general conclusion by suggesting that probably all other theoretical models and works in social sciences have their own political implications and dimensions.[35] We can even say, in this sense, that this chapter, this theoretical work, is also political. While showing the antagonism, criticism, and political trends inherent in the theoretical model of social fluidity, it invites us to get involved in this model and participate in that kind of (fluid) politics.

Notes

1 For example, the Actor-Network Theory helps us to regard network relationships and agent positions as part and consequence of the ongoing assemblage (or fluidity) in an increasingly complex social reality.
2 See García Selgas (2001, 2002, 2003).
3 Like structural functionalism at the dawn of the twentieth century with its scientific will and its balance between socialist and liberal values; or the microsociologies (ethnomethodology, symbolic interactionalism, etc.) of the 1960s with their commitment to understanding real social practices and their involvement in the decade's new social movements.
4 However, there are important contributions like Mol's (1999) notion of "ontological politics" (i.e., different, competing versions of reality implemented by different techno-scientific practices), and Harding and Haraway's idea that we remain within the scientific project knowing that we are practicing politics by different means. It is also important to mention that the etymology of "theory" brings in the idea of vision, of what we can see: the possibilities that are visible and, as such, available.
5 Castells (1996) with his society of information fluxes, Mol and Law (1994) with their fluid social space, Bauman (2000) with his liquid society, and Semprini (2003) with his society of flux.
6 Similar research may reveal that the substantialist model is linked to a kind of Hobbesian view of politics as an arrangement of agreement (among individuals) and (state) control, that reduces the political realm to government (like Weber), while the formalist model relates the political to the supra-individual forces or the dynamics of development and the closure of some totality, as in Marx and Engel's *Communist Manifesto.* If these results emerged from research, we would have a much stronger argument for the internal connections between theoretical models of social reality and politics.
7 See, e.g., M. Castells in his trilogy *The Information Age* (1996).
8 Other artworks at the same exhibition produced similar results (e.g., Simryn Gils denouncing the complexities of colonialism and advocating hybrid social beings, and Nedko Solakov showing the involvement of many people, even himself, in Eastern European state surveillance systems, while moving towards internal criticism and proof of a "two-fold concept of truth"). See references in *Documenta* (2007: 130–131, 236–237, and 250–251).
9 It should be remembered here that Ethnomethodology, Symbolic Interactionalism, and the Actor-Network Theory have already conducted important research into the regular (micro) practices which, while sustaining social institutions, fill them with conflicts, controversies, and antagonisms.
10 Here he develops Debord's notion of "Society of Spectacle" and Deleuze and Foucault's "Society of Control".
11 What else have celebrities like Sharon Stone, Angelina Jolie, Bono, and Pelé been doing at the recent meetings of the powerful Davos group besides capturing our attention? And we continue to watch the stars.
12 In this sense, there is no relevant difference between the two paradigmatic cases: the liberal version of the French Revolution, storming the Bastille, and the Marxist cause of the Soviet Revolution, raiding the Winter Palace.
13 Authors such as Foucault, Deleuze, Guattari, Negri, and Hardt also speak of a transition from a disciplinary society, where domination is exercised through intermediate, closed, concrete institutions (schools, jails, hospitals, etc.), to a society of control, where domination and hierarchical differentiation pervade all material and symbolic social spaces. For them, most political and institutionalized practices in representative democracies are performed solely in the interests of

their own reproduction and the expansion of silent consensus, which turn them into conservative practices within this logic of control, while they see most forms of politics of identity as excluding others. They thus locate the potential for resistance to this pervasive control in the questioning, swaying, and alteration of these forms of political activity, seeking a constantly open, undecided, fluid political situation. This conclusion denies a solid, stable political (and critical) body, and calls for an open, multiple process with no general solutions or indications. It is perfectly compatible with the fluidity model. However, the sort of dual perspective underlying these concepts (arborescent versus rhizomatic or state versus nomadic, in Deleuze's terms), which ultimately makes the practices of resistance a case of willingness, are not compatible. According to the fluidity perspective, open, unstable, fluid political resistance is not just something we seek but something we already find in (fluid) social reality, including dominant (fluid) dynamics.

14 In an oversimplified way of expressing the historical transition, one could speak of an initial transition from use-value to exchange-value (with money as a universal measure of value) and a second one from exchange-value to sign-value (when the monetary value depends on what social agents believe, understand, or expect).

15 Bauman (2000: 131–135) links this question about agency to the post-industrial (and post-Hiroshima) loss of faith in human progress, which means, among many other things, a loss of faith in the idea that time is on our side and in the modern belief that the capacity for change and improvement is in our hands, as well as the regularization and privatization of progress.

16 To some extent, this is what Fukuyama means when he speaks of "the end of history".

17 Other political agents involved in this issue such as the Catholic Church, political parties, urban middle classes, etc. basically regarded the law as an important step towards a more secular way of life, as one step out of the traditional moral order. This understanding is probably also partly true due to the multiplicity of a fluid ontology (Mol).

18 Something of this sort may have been learned by the Spanish Catholic Church when one of its main claims was that unions between homosexuals should be called something other than marriage. They were right: it was more that a simple law change. Two points about the Catholic hierarchy in Spain are worth mentioning: it is currently extremely conservative and still has a powerful moral influence, primarily but not exclusively among right-wing voters.

19 There is not only dissolution of neat distinctions between domination spaces and the realms or mechanisms of resistance within the fluid theoretical model, but also a shift of social and political agency from human hands into the capabilities of other (social) beings and objects. The "cyborg" notion in Haraway's work is a fine example of this movement towards a multiple, heterogeneous political agency.

20 The fluidity model is generally averse to any kind of ontological dichotomy between social system and social agency, e.g., empire vs. multitudes, discipline vs. control, biopower vs. bioproduction, etc.

21 Negri and Hardt (2004) speak of "multitude" as an open and flexible network allowing the expression of differences but producing a political unity, a common reference around general social wealth, a new social body.

22 Basic references include Mol's notion of "ontological multiplicity" and Sthratern's "partial connections".

23 Arditi and Hequembourg (1999: 62). Their analysis of debates in US gay circles about the best tactics of resistance (assimilation versus affirmed desire) has inspired the argument that the accumulation of partial modifications is what we need for an overall transformation of the field of power and the coherence of inconsistent oppositional agents.

24 From the social fluidity perspective, "partial modifications" are not any kind of renewed reformist policies such as most social democrat politics are. They are more like changing sensibilities, discourses, as well as power and material relationships, because (fluid) social time–spaces overflow closed national territories with information, capital, and people, and most (fluid) agencies are parts and effects of changing material–semiotic fluxes.

25 The very first sentence of the *Tao te ching* says: *"dao ke dao fei chang dao"* which may be translated as "The flow that can be followed is the flow in constant transformation".

26 The notion of "criticism" has been a core concept in the three separate spheres of modern thought: theoretical, scientific, and epistemological (with notions like verification, testing, etc.); practical, ethical, and political (with notions like utopia, common good, natural law, etc.); and aesthetic, artistic, and literary (interpretation and judgement of the merits of a work of art). The concept of "objective criticism" was a central part of basic modern paradigms: liberal (from Comte to Popper, where it had a cognitive and formal nature) or socialist (from Marx to Adorno, where it was primarily social and material).

27 Assumed as part of the emancipatory aim of total autonomy is the idea that the only constituent parts of society are human beings with their actions and institutions, which is an important idea in other theoretical models of social reality.

28 It is in this sense that Latour calls for an "open parliament" that also includes the voices of non-human social agents, and Haraway calls for a politics of cohabitation and co-evolution with "companion species". A beautiful connection to environmentalist, political stances may be found here.

29 The fluidity model is, incidentally, far from a regulative idea of final reconciliation (like paradise) that has been operating as a vanishing call to a constant escape from past and present. Unstable multiplicity and the other basic features of a fluid social world are much closer to the regulative idea of promiscuity in different and possible presents.

30 In all physical fluids, resistance is a basic internal (loss of mobility or variability, stabilization) and external (resistance to bodies plugging into them) feature that works in a very complex way (e.g., water resistance to a ship is the condition for its flotation).

31 If some political forms and actions are made visible, it is not only a consequence of our own making or naming (as some constructivists believe) but because the necessary conditions, in which objects and other non-human agents take part, are given. It is a co-construction.

32 We thus have an internal confirmation of the political nature of scientific discourses: by claiming "the truth" they become disputed contributions to enact one possible version – and not another – of what is going on.

33 See Mol (1999: 79–86).

34 Given additional time–space, this conclusion could be made stronger and more compelling by showing that the changes undergone by many contemporary societies drive them into a state of fluidity.

35 I am not suggesting the determination of politics by social theories nor the other way around, but a mutual and internal conditioning within an ongoing, common historical process.

References

Arditi, J. and A. Hequembourg (1999), "Modificaciones parciales: discursos de resistencia de gays y lesbianas en Estados Unidos", *Política y Sociedad*, 30.

Bauman, Z. (2000), *Liquid Modernity*, Cambridge: Polity Press.

Castells, M. (1996), *The Information Age. Vol. 1: The Rise of the Network Society*, Cambridge, MA: Blackwell.
De Laet, M. and A. Mol (2000), "The Zimbabwe Bush Pump: Mechanics of a Fluid Technology", *Social Studies of Science*, 30(2).
García Selgas, F. (2001), "Preámbulo para una ontología de la fluidez social", *Atenea Digital*, 1. www.blues.uab.es/athenea.
—— (2002), "De la sociedad de la información a la fluidez social", in J.M. García Blanco & P. Navarro (eds), *¿Más allá de la Modernidad?*, Madrid: CIS.
—— (2003), "Para una ontología política de la fluidez social: el desbordamiento de los constructivismos", *Política y Sociedad*, 40.
Haraway, D. (1992), "The Promise of Monsters: A Regenerative Politics for Inappropiate/d Others", in L. Grossberg, C. Nelson & P. Treichler (eds), *Cultural Studies*, London: Routledge.
Laclau, E. and C. Mouffe (1985), *Hegemony and Socialist Strategy: Towards a Radical Democratic Politics*, London: Verso.
Latour, B. (1991), "Technology is Society Made Durable", in J. Law (ed.) *A Sociology of Monsters: Essays on Power, Technology and Domination*, London: Routledge.
Law, J. (1994), *Organizing Modernity*, Oxford: Blackwell.
Mol, A. (1999), "Ontological Politics. A Word and Some Questions", in J. Law & J. Hassard (eds), *Actor Network Theory*, Oxford: Blackwell.
Mol, A. and J. Law (1994), "Regions, Networks, and Fluids: Anaemia and Social Topology", *Social Studies in Science*, 24: 641–671.
Mouffe, C. (1997), *The Return of the Political*, London: Verso.
Negri, A. and M. Hardt (2004), *Multitude: War and Democracy in the Age of Empire*, New York: The Penguin Press.
Semprini, A. (2003), *La società di flusso*, Milan: Franco Angeli.

8 Collateral realities

John Law[1]

Introduction

Welcome to the world of collateral realities.

Collateral realities are realities that get done incidentally, and along the way. They are realities that get done, for the most part, unintentionally. They are realities that may be obnoxious. Importantly, they are realities that could be different. It follows that they are realities that are through and through political.

This chapter explores some of the ways in which realities including collateral realities get done. Note the verb. Not *known*, but *done*. Here is the first blockage. To talk of doing realities is to push outside the comfortable envelope of Euro-American common-sense realism. It takes us into a world of serious performativity.

So what is 'Euro-American common-sense realism'? There are whole libraries on this, but here is a gesture. First, it tells us – it assumes – that there is *a reality out there*. Second, it tells us that whatever is out there is largely *independent* of our actions. (A qualification: it is obvious that our actions sometimes influence reality.) Third, it tells us that whatever is out there substantially *precedes* our actions or attempts to know it. Fourth, it assumes that whatever is out there is *definite* in form. Fifth, it takes it for granted that there is a single reality, that it is *singular*. And, sixth, probably (perhaps less certainly) it assumes this reality to be *coherent*.[2]

We may debate the specificities, but if we take performativity seriously then most of these assumptions need to be undone. Only a stripped-down version of the first (call this 'primitive out-thereness') remains. If we think performatively, then reality is not assumed to be independent, priori, definite, singular or coherent. Rather the logic is turned upside down. If reality *appears* (as it usually does) to be independent, prior, definite, singular or coherent then this is because it is being *done* that way. Indeed, these attributes or assumptions become examples, among others, of collateral realities.

But what is it, 'to do'? Where are the collateral realities being done? The response is that they are done in *practices*. Practices enact realities including collateral realities. This means that if we want to understand how realities are done or to explore their politics, then we have to attend carefully to practices and ask how they work. Libraries have been written on this topic too, so I simply offer another gesture. For my purposes, practices are *detectable and somewhat ordered sets of material–semiotic relations.*[3] To study practices is therefore to undertake the analytical and empirical task of exploring possible patterns of relations, and how it is that these get assembled in particular locations. It is to treat the real as whatever it is that is being assembled, materially and semiotically in a scene of analytical interest. Realities, objects, subjects, materials, and meanings, whatever form they take these are all explored as an effect of the relations that are assembling and doing them. Practices, then, *are* assemblages of relations. Those assemblages *do* realities. Realities, including the incidental collateral realities, *are inseparable from the patterning juxtapositions of practices.*

There is an immediate methodological consequence. We need to proceed *empirically.* If we are to do philosophy, metaphysics, politics, or explore the character of knowledge, we cannot do this in the abstract. We cannot work 'in general', because there is no 'in general'. All there is are: specific sites and their practices, and then the specificities of those practices. So philosophy becomes empirical.[4] Abstraction is always done in some practice or other. As, to be sure, are collateral realities.

For this reason, in what follows I work empirically and attend to specificities. My interest is in how realities (and representations of realities) are assembled in material–semiotic relations at a particular place, moment, and occasion. The place I talk about is a lecture hall in a research institute in Berlin. The moment is a meeting that took place in that hall in May 2007. The occasion was a stakeholders' meeting of a programme called Welfare Quality®. The latter was a large-scale EU-funded Framework 6 Programme on farm animal welfare. It was about farm animal welfare and included: animal science research, a major social science contribution on what European consumers think about farm animal welfare, and the development of a series of far-reaching proposals for monitoring and reporting on that welfare. In May 2007, the moment of this meeting, the programme was half-way through its five-year life, and those caught up in it, including animal scientists, social scientists, animal welfare NGOs, the food trade, and farmers, were debating the form that the ultimate recommendations might take. In short, a great deal was at stake for many of the participants.

Why do I focus on this meeting? One response is straightforward. As I have just said, much was at stake: this was an overtly *political* meeting. Second, however, and as a part of this, I am interested in it because realities were also being negotiated. What is a *farm animal*? What is a *consumer*? What, for that matter, is *welfare*? At this meeting these issues were all being contested. In other words, the meeting was not simply about politics as this is conventionally understood, but also involved a 'politics of the real'. There were struggles between different versions of reality: this was, in short, a moment of *ontological politics*. Third, the reals at stake were sometimes explicit, but very often they were not. *Collateral realities* were being done too and I am particularly interested in exploring some of these. And then, finally and crucially, I am interested in the character of the reality work being done in a *meeting*. We tend to think of laboratories or social science surveys as locations where the character of realities such as animals or people is determined, but realities are done in meetings too. Like laboratories, these are assemblages framed in particular ways. Like surveys, they are sets of practices, both patterned and patterning, where different ontological politics and different collateral realities are routinely done. So this chapter is also an exploration of how realities emerge from meetings.

Lecture

The scene, then, is a lecture hall in a research institute in Berlin. Here's what it looks like from the back[5]

It is the morning of the first day of the meeting, and the room is nearly full.[6]

The audience is listening to a talk. Here's the first PowerPoint of that talk.[7]

And here are the opening paragraphs of the abstract for the talk (everyone has a copy of this).[8]

Welfare Quality®: context, progress and aims

Harry Blokhuis, Animal Sciences Group, Wageningen University and Research Centre, The Netherlands

Some five years ago we started to formulate the first aims and goals of what became the largest piece of integrated research work yet carried out in animal welfare in Europe. The Welfare Quality® project has now been running for three years and it has been an exciting and very productive time. We are also making a global impact. For example, the Welfare Quality® consortium was recently extended to include four Latin American partners, so we now have a total of 44 partners from 17 countries.
Although the originally formulated goals have evolved as results emerged and as opportunities arose, the main aims still stand:

- to develop practical strategies/measures to improve animal welfare,
- to develop a European standard for the assessment of animal welfare ,
- to develop a European animal welfare information standard,
- to integrate and interrelate the most appropriate specialist expertise in the multidisciplinary field of animal welfare in Europe.

Those who are present have been told that the speaker is Professor Harry Blokhuis. If we just depend on Blokhuis' own materials we can see that he is talking about a large integrated European research project on farm animal welfare. We know or can infer from these exhibits that this is called Welfare Quality®, which (its logo tells us) is about 'Science and society improving animal welfare'. We can guess (look at the Sixth European Framework Program logo) that it is being funded by the European Commission. We know from the text that the project extends beyond Europe. We know that it is intended to develop practical strategies and measures for assessing and improving animal welfare, including the creation of an 'information standard'. And we know that it is intended (I quote from the abstract above) 'to integrate and interrelate the most appropriate specialist expertise in the multidisciplinary field of animal welfare in Europe'. Now look at this (PowerPoint number six of Blokhuis' presentation).[9]

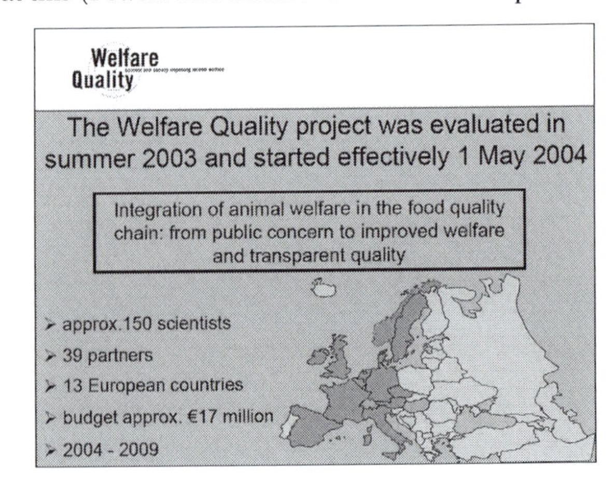

We also learn about the budget (around €17 million), the span of the project (2004–2009), and something about its spread (150 scientists and thirty-nine partners across thirteen European countries).

Now note that unless you already knew of the project, then the way in which I have represented it above is all that you know about it right now. The reason I say this is that I want to draw your attention to the *performativity* implied in (your reading of) my description. To put it differently, everything else being equal (probably it is not), for the moment what I have described and shown *is* what Welfare Quality® is for you. My text and your reading practices have assembled a putative Welfare Quality® reality, at least for the moment. This is the performativity of practice at work. I am suggesting that my textual and your reading practices are together assembling a putative reality.

Now bracket this reflexivity away, and attend instead to Blokhuis' talk and its reception in the hall. I want to say that this too is a set of practices that is assembling a putative reality: that it is *doing* a possible reality. A clarification. In saying this I am *not* criticising or trivialising either the talk or its reception. My interest is quite different. It is to ask *how* these talking and meeting practices work to assemble a putative reality. But if we are to do this then we have to teach ourselves to *see* the work being done by the PowerPoints and the abstracts. We need to find ways of making this work visible. We need to resist the propensity to treat these texts as transparent, self-evident, or uninteresting windows on a pre-given world.

It may help us to do this by looking at a representation that is not particularly clear (see figure below). This comes from my notes on Blokhuis' talk. The fact that they are not very convincing serves to remind us that *all* representations – notes, PowerPoints, photographs – are the product of *practices*. Here, for instance, I was writing frantically, but Blokhuis was not. He had written a careful abstract and well-crafted series of Power Points. But (here is the important point) the *principle* at work is similar in both. Both note-taking and talk-preparing are more or less ordered practices. Both generate representations that depict realities. Both, I am saying, are helping to *assemble* putative realities. And since those realities are being done in particular ways, at least in principle this also implies that they could have been assembled differently. This is why I am saying that we are watching a form of politics, *ontological* politics. For while it is more or less received wisdom that representations are not more or less clear windows on reality, but shape, form and diffract reality,[10] I am making a stronger claim. If, performatively, representations do realities in practice, then those realities might have been done differently. We find ourselves in the realm of politics.

2. No sanctions to reduce density
3. Purpose lessons, loss. (Only indicator allowed: mortality)

Harry Blockhuis
2 lines of info
1. Consumers – bothered about welfare, but lacked informati
lack of info
2. Biological science
Combine scientific expertise from soc sci + animal scien
2. Project
-2004-9 150 scientists
2006-Latin American partners

More on Blokhuis' lecture

So how does Blokhuis' talk work?

Here are some of the processes upon which it depends: *selection, juxtaposition, deletion, ranking,* and *framing.*[11] Like my field notes, the photos, and the abstract, every PowerPoint operates in one and probably most of these modalities. For instance, we have already seen the figure below.[12]

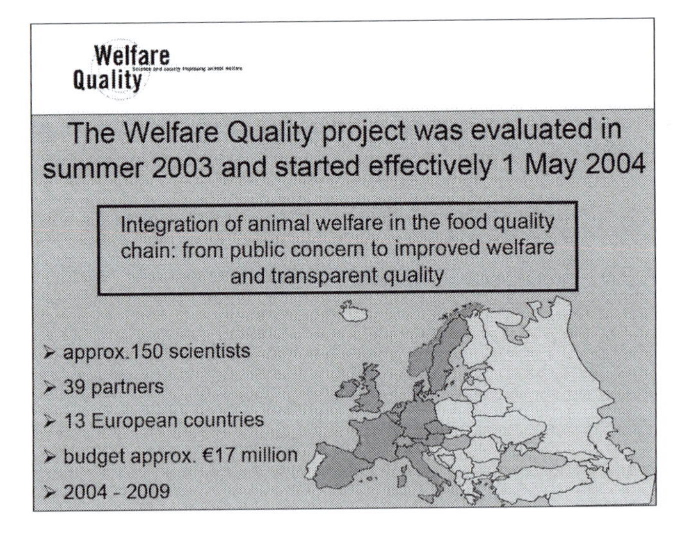

Look the juxtaposition and the framing at work here. This does Welfare Quality as or in a set of relations that are simultaneously *financial* (there is a budget), *geographical* (there are thirteen European countries and a map), *scientific* (or at least there are scientists), and *chronological* (there are dates and a time-span). Performatively, for the moment this is what Welfare Quality *is*: a juxtaposition of selected elements. Let me add that it is also *teleological.* Welfare Quality® is being done as a project with a purpose: 'Integration of animal welfare in the food quality chain: from public concern to improved welfare and transparent quality.'[13] A similar teleological reality is being done in other PowerPoints (see figure below).[14] Milestones belong both to projects and chronology, and both are being done here.

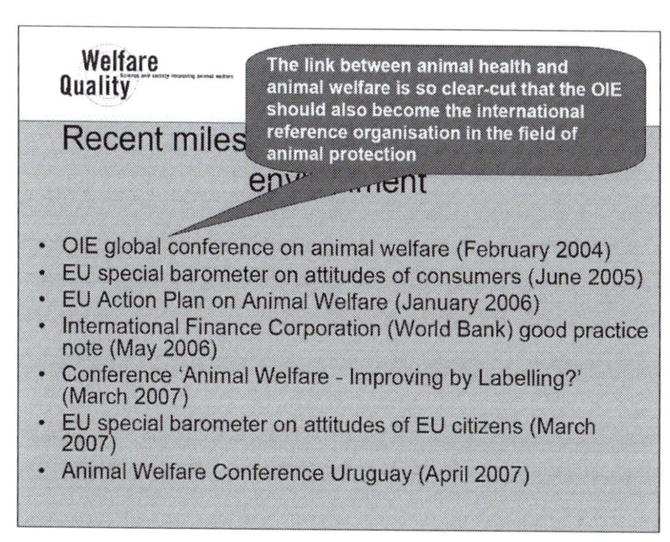

So there are relations, juxtapositions and framings. There are also rankings. Some elements are important enough to appear on the PowerPoint (certain dates, budgets, geographies, scientists and partners), whereas others are not and fall off the edge. So what is being deleted? In principle the answer is: almost everything. Indeed it could be no other way: the ramifying complexities of the whole world cannot be included on a single PowerPoint. But one way of rendering this question tractable is to compare and contrast different depictions. Look, for instance, at this figure.[15]

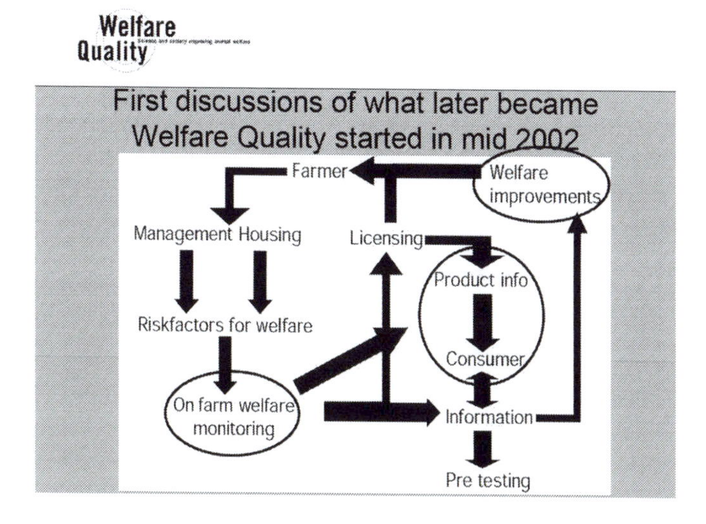

Here we see the Welfare Quality® framing again, and a title enacting the project in chronological time (2002); but otherwise the principles at work are different. The criteria for selecting, juxtaposing and ranking have all changed. This is a cybernetic world made up of interconnections with feedback loops between actors that are thereby rendered significant. Farmers and consumers interact with a heterogeneous population of other elements including the quasi-legal ('Licensing'), the organisational ('Management'), the material ('Housing'), the procedural ('On farm welfare monitoring'), what one might think of as ethical or political realities ('Welfare improvements'), and abstractions ('Information'). Farmers are being defined by their *relations*. So they are linked with welfare improvements, licensing, management and housing. Consumers are defined by their connections with production information, on-farm welfare monitoring, and (more generally) information. The material–semiotic relations are being laid out visually. A system or network world is being done while geography, finance, and to a lesser extent chronology have all disappeared.

See below for a further PowerPoint from Blokhuis' talk.[16]

Look at the row of people-figures. Presumably these are animal scientists and social scientists. They are combining their essential scientific expertise in the WQ programme in order to simultaneously understand (I impute) *consumers* (like the man with the shopping trolley in the photo in the left) and *animals* (the piglets in the second photo on the right). But we have moved on from a cybernetic world. The ordering has changed. Consumers 'belong' to social sciences, and animals to animal science. Here system, chronology, geography and finance have been deleted. The principles for selecting, juxtaposing and ranking are quite different. What is being assembled is a world in which there is an intellectual and social division of labour between the domains of animal and social science. This, it turns out, is not hierarchical: these two domains cooperate by together contributing to the 'Welfare Quality Consortium' and ultimately to the Sixth Framework Project.

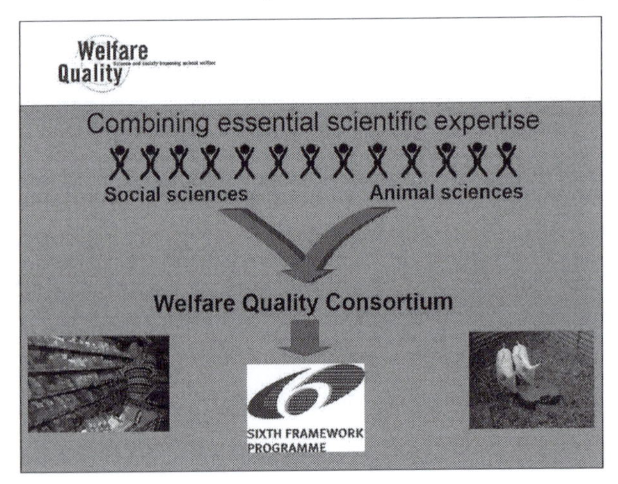

So what do we learn if we attend to the PowerPoints in this way? What happens if we *see* them and the work they are doing, and manage to treat them as part of – and an expression of – practice, rather than as more or less transparent windows on a pre-given reality? The answer comes in three parts:

- The first is that it becomes possible to explore the character of their performativity. The PowerPoints do work, and as I have tried to show, they do this by selecting, juxtaposing, deleting and ranking. All in all, they work by framing. This is a *methodological* point. We need to overcome the obviousness of representations if we are to understand how they work.
- Second, we discover that the way in which they work is quite startlingly varied, for it turns out that quite *different* Welfare Quality® realities are being done at different moments. As I have shown, the first PowerPoint does the programme as some kind of *genealogy*, as a teleological project with its roots and origins spread through layers of time. The second performs it quite differently as a heterogeneous cybernetic *system*: here Welfare Quality is being done as a set of feedback loops that are indifferently social, political, animal, industrial, and normative. And the third does Welfare Quality differently yet again. Here it becomes a form of professional cooperation, an expression of the division of scientific labour. *Interdisciplinarity* is being done.[17]
- Genealogy, system, and interdisciplinarity: if we read these representations as enactments by asking *how* they work we also discover that in five minutes in a single lecture hall this project has been done in three quite different ways. Let me remind you that this is not a complaint. On the contrary, attention to the specificities of practice and its enactments usually uncovers difference, and suggests that non-coherence is a chronic condition.[18] It may well that such multiplicity is a necessary condition for institutional survival.[19] But if this performative way of thinking shows that reals are done in multiple ways, then it also suggests that at least in principle those realities – or the balance between them – could be different. And this is my third point. To attend to the specificities of practice leads us to the possibility of an *ontological politics.* At the same time it allows us to explore the enactments of collateral realities. For what are genealogy, system and division of labour if they are not collateral realities, versions of the social that are being done quietly, incidentally, and along the way?

Practices all the way down

Let me briefly revisit the question of realism.

I have suggested that common-sense realism tells us that realities are independent, and prior to practices. They are also taken to be definite, singular and coherent. In this way of thinking, depictions (talks, charts, PowerPoints) more or less adequately represent those realities.

Comment. Philosophers who worry about the adequacy or otherwise of representation are often called *epistemologists.* Practitioners who attend to this are sometimes called *methodologists.* Both endeavours have produced large libraries. There are many specificities, but if we stick with the methodologists, then we know that they worry about technical adequacy. The assumption is that good techniques produce satisfactory representations of reality. What follows? One implication that I have already touched on is that techniques themselves become essentially uninteresting. This is because when they are working properly they are transparent. In this way of thinking they do not distort realities, but merely transmit them. In short, good methods are like a window on reality.[20] This means that unless something has gone wrong they can be ignored. As is clear, I have been arguing against this. No representation, I have been saying, is actually transparent.

Now look at the figure below.[21] These words come from Blokhuis' abstract. 'Recent surveys … confirm that animal welfare is an issue of considerable significance for European consumers.' '*Recent surveys confirm*' (my emphasis). The words (appear to) open a small window on to reality. At the same time (this is a part of the realist trick) the methods for making that window have been more or less erased. All we get to see is a European reality composed of European consumers and citizens. 'Eurobarometer 2005' is being done here as a source of representational authority, but at the same time it is essentially uninteresting. We do not need to know about the methods involved. This is presumably because it may be assumed that the technique mechanically discovers the nature of a particular European consumer- and citizen-reality and then reports on it. Put that performatively. Survey research is being *done* here as a window on a specific reality.

> was produced. In other words, animal welfare is an important attribute of an overall 'food quality concept'. Recent surveys carried out by the European Commission (e.g. Eurobarometer, 2005[1]) as well as studies within Welfare Quality®, confirm that animal welfare is an issue of considerable significance for European consumers and that European citizens show a strong commitment to animal welfare.

Now look at the figure below.[22] I have moved to the next talk of the day. This is by three Welfare Quality social scientists, Unni Kjærnes, Emma Roe and Bettina Bock, and it is more specific. We learn (for instance) that the French public is more worried about farm animal welfare than the Swedish, and in all seven countries the public worries most about the welfare of chickens. Once again the methods and how they work have been deleted. They have been done as essentially uninteresting, a means to an end, another window on reality.

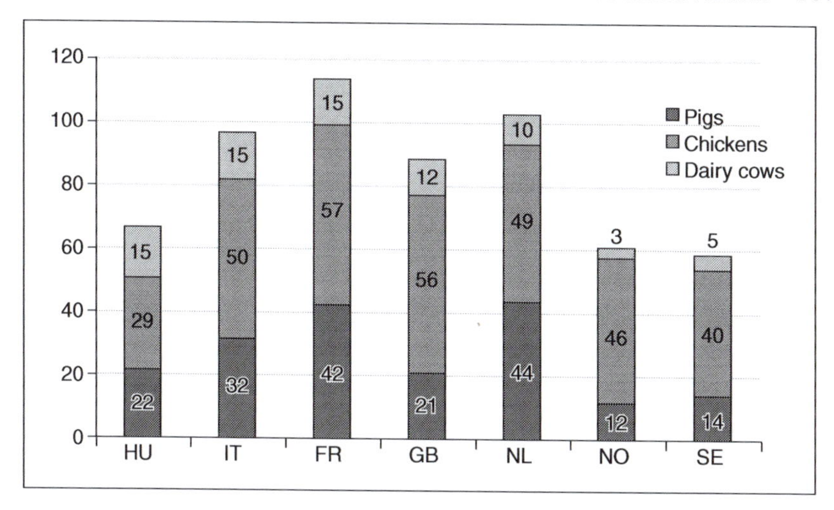

One consequence of this is that the assumptions of common-sense realism are being redone. We have moved from European realities in the form of 'consumers' and 'citizens' in Blokhuis' talk, to another set of possibly related realities, this time in the form of 'publics'. It is now the latter that are being enacted as prior to and independent of the research. They are also being done as having definite opinions, and collectively they have been rendered singular and coherent. As we can see, it is possible to talk of 'public responses' in, for instance, Hungary. Such are the kinds of reality work being done in Berlin in these presentations. But there is implicit work being done too. So, for instance, realities such as nation-states are *implied* in claims about the views or attitudes of groups of people in Hungary or France. In other words, the nation-state is a collateral reality being carried along and enacted in the wake of explicit research findings. The survey does not set out to demonstrate the existence of the nation state, that is not its point at all, but it does this quietly and therefore all the more effectively.

The argument works for individuals as well as for collectivities. We can see how this works by looking, for instance, at survey research methods. This is because like PowerPoints, surveys also delete, select, juxtapose, rank, and frame to enact a version of the real. Thus in one of the Welfare Quality® commissioned surveys, people were phoned and asked: 'Thinking of farm animal welfare in general, how important is this issue for you on a scale of 1 to 5, where 1 is not at all important and 5 is very important?' Critics of survey research would say that this deletes important ambivalences, uncertainties and situational complexities: that people are not *really* like that. They might add that complexities are better explored in, say, focus groups.[23] This may be right, but it also misses the performative point. This is that *both* focus groups *and* surveys delete, select, juxtapose, rank, and frame realities. Both, that is, *enact* reals and (here is another collateral reality) they enact people in particular and distinctive ways. For if focus groups may be understood as producing talk about situated political and community positions and debates, then a survey question of the kind that I have just quoted assumes that individuals may be understood as containers of attitudes that are somewhat stable and behaviourally relevant. Indeed, it enacts them that way. In short, models of the individual as well as the collectivity are being performed – and if the findings derived from that model are taken seriously by audiences such as those in the Berlin hall, then those models count as another collateral reality.[24]

It is important to add that this is not just the case for the social world and its social sciences. It applies just as much to natural science and the natural world. Look, for instance, at the figure below.[25] These are the opening lines of the abstract of the third talk of the Berlin morning by animal scientist Isabelle Veissier and sociologist Adrian Evans. 'To date' it says, 'no unique measure of welfare exists … welfare is a multidimensional concept. It comprises both physical and mental health … and includes several aspects such as physical comfort, absence of hunger, diseases, or injuries.' Here is one of their PowerPoints (see the following figure).[26] A few minutes earlier a *social* and consumer reality was being done, but now it is the turn of the *animal* and the natural world. Welfare – the animal-with-welfare – is being staged here in four large categories: 'Good feeding', 'Good housing', 'Good health', and 'Appropriate behaviour', which are then broken down into twelve sub-criteria. Common-sense realism is hard at work again. For the moment the animal-with-welfare *is* this way. The PowerPoint *is* a window on to this reality. Footnotes and references aside, the methods and the practices for doing this animal-with-welfare have been deleted. As with the survey research that staged the consumer, *how* they work is being treated as essentially uninteresting. It is the reality they end up describing and enacting that is the focus of attention. The methods themselves, and the assumptions that they enact, are erased. In short, there are collateral realities being done here too. So what do they look like?

Welfare®
Quality
Science and society improving animal welfare

Rationale behind the Welfare Quality® assessment of animal welfare

Isabelle Veissier, INRA, UR1213 Herbivores, F-63122 Saint Genes Champanelle, France
Adrian Evans, School of City and Regional Planning, Cardiff University, United Kingdom

Ensuring the welfare of animals that produce products for human consumption requires means to reliably assess animal welfare and to inform - in a standardised way - producers, retailers, consumers, animal protectors, and a range of other citizens. To date no unique measure of welfare exists. This is essentially due to the fact that welfare is a multidimensional concept. It comprises both physical and mental health (Dawkins 2006; Webster 2005) and includes several aspects such as physical comfort, absence of hunger, diseases, or injuries etc. (Farm animal welfare Council 1992). The importance attributed to different aspects of animal welfare may also vary between people (see Fraser 1995).

Welfare® Quality Science and society improving animal welfare

WelfareQuality® list of criteria

Welfare criteria	Welfare subcriteria (principles)
Good feeding	1. Absence of prolonged hunger
	2. Absence of prolonged thirst
Good housing	3. Comfort around resting
	4. Thermal comfort
	5. Ease of Movement
Good health	6. Absence of injuries
	7. Absence of disease
	8. Absence of pain induced by management procedures
Appropriate behaviour	9. Expression of social behaviours
	10. Expression of other behaviours
	11. Good human-animal relationship
	12. Absence of general fear

(Botreau et al., 2007; WQ fact sheet Veissier & Miele 2007)

As with the social sciences, it is methodologically helpful to look for differences between practices within animal science, and to search for contrasts in the ways that these delete, select, juxtapose, rank and frame their reality, the animal-with-welfare. Look, for instance, at the figure below, which comes from the scientific literature:[27] Here *poor welfare is not necessarily the same as suffering* but/and poor welfare can be measured *objectively* ('reduced life expectancy, impaired growth, impaired reproduction, body damage, disease'). What is being implied? The answer is that the animal is being done as a *body*. This is a body endowed with *clinical and endocrinological attributes* that may be measured by the instruments of animal science. As a part of this, emotions and experiences are made not to count. But this is only one possibility. Look at the figure following this.[28] Here we have a different scientific article, a different author, and a different set of practices. For here there is talk of 'positive states'. These may be difficult to study, but in this practice animals enjoy 'presumably pleasurable activities, such as play and exploration'. '[C]ats', we learn, '*derive pleasure* from being stroked'. Here, then, though bodily states are not being deleted, suffering and pleasure are also important. The animal is being done differently. The animal is not just a body. It is not just an *object*, but it becomes a *subject* too. The framing and the mode of deletion here are both different.

ANIMAL WELFARE: CONCEPTS AND MEASUREMENT[1,2]

D. M. Broom

Cambridge University[3], Cambridge, CB3 OES, United Kingdom

ABSTRACT

The term "welfare" refers to the state of an individual in relation to its environment, and this can be measured. Both failure to cope with the environment and difficulty in coping are indicators of poor welfare. Suffering and poor welfare often occur together, but welfare can be poor without suffering and welfare should not be defined solely in terms of subjective experiences. The situations that result in poor welfare are reviewed in this study with special reference to those in which an individual lacks control over interactions with its environment. The indicators of poor welfare include the following: reduced life expectancy, impaired growth, impaired reproduction, body damage, disease, immunosuppression, adrenal activity, behavior anomalies, and self-narcotization. The uses of measures of responsiveness, stereotypies, and animal preferences in welfare assessment are discussed. The need to make direct measurement of poor welfare as well as to use sophisticated studies of animal preferences is emphasized.

Key Words: Welfare, Responses, Pain, Preference Tests, Behavior

J. Anim. Sci. 1991. 69:4167–4175

Third, well-being implies that the animals should have positive experiences, such as comfort and contentment, and freedom to engage in presumably pleasurable activities, such as play and exploration. This aspect is likely to cause the most controversy. Animal welfare scientists recognize the difficulty of studying positive states such as contentment (Fraser and Broom, 1990). Scientists in the behaviorist tradition may dismiss such states as outside the realm of science or deny that animals experience such states at all (Rollin, 1990). On the other hand, many people who have worked closely with animals consider it ludicrous to deny, for example, that dogs *enjoy* playing or that cats *derive pleasure* from being stroked. For such people, depriving animals of pleasure is one of the fundamental issues in animal well-being (see Harrison, 1964). Hence, despite the difficulty of studying states such as contentment, we cannot realistically exclude them from our criteria for well-being.

Finally, note that both versions of the animal turn up in the list of Welfare Quality criteria (the figure below).[29] Criterion number 4, 'Appropriate behaviour', welfare principle number 10, 'Expression of other behaviours': 'Animals should have the possibility of expressing other intuitively desirable natural behaviours, such as exploration and play.' After looking for different practices of enactment within the animal science literature it now becomes clear that both versions of the animal-with-welfare – the animal as object and the animal as subject – are being done in Veissier's and Evans' talk, and in the putative reality being assembled by Welfare Quality®.

			harmful, social behaviours.
Appropriate behaviour	10.	Expression of other behaviours	Animals should have the possibility of expressing other intuitively desirable natural behaviours, such as exploration and play
	11.	Good human-animal relationship	Good Human-animal relationships are beneficial to

Why is this important? The answer is partly methodological and partly political, and it also has to do with the distinction between natural science and social science. It is often suggested that the latter are different in kind. No doubt for certain purposes this is right. However, for present purposes it is quite wrong. This is because both, I am arguing, may be understood as *sets of practices.* Indeed, they may be understood as sets of practices all the way down. Whether we look at social science reports of reality or those coming from natural science, once we start to turn up the magnification we quickly find that there is not an independent, prior, definite, singular and coherent real out there upon which the various reports of reality are based. Instead, what we find is more practices doing reals. And more practices. And yet more practices. And since we also find as we turn up the magnification that there are different practices within each domain, we also discover multiplicities – different versions of the animal as well as different versions of the person. In short, I am saying that performativity is everywhere in natural science as well as in social science. The implication is that the character of the real is as open to debate in nature as it is in society. An ontological politics is possible – and collateral realities may be found – in both.

Collateral realities

Here is the argument.

First, attend to *practices.* Look to see what is being done. In particular, attend empirically to *how* it is being done: how the relations are being assembled and ordered to produce objects, subjects and appropriate locations. Second, wash away the assumption that there is a reality out there beyond practice that is independent, definite, singular, coherent, and prior to that practice. Ask, instead, how it is that such a world is *done* in practice, and how it manages to hold steady. Third, ask how this process works to *delete* the way in which this sense of a definite exterior world is being done, to wash away the practices and turn representations into windows on the world. Fourth, remember that wherever you look whether this is a meeting hall, a talk, a laboratory or a survey, there is *no escape from practice.* It is practices all the way down, contested or otherwise. Fifth, look for the gaps, the aporias and the tensions between the practices and their realities – for if you go looking for *differences* you will discover them.

These are the steps to follow if we are to attend well to practices, specificities, processes and materialities. They are also the steps that are needed if we are to undo the metaphysics of common-sense realism. Is reality destiny? Common-sense realism says yes. It suggests that while we may try to engineer the world and influence it, in the end the world is arranged in the way that it is: fixed more or less, definite more or less, and singular, coherent and outside practice. The move to performativity says no. It allows us to ask questions about realities that are simultaneously analytical and political. We may begin to ask how they are done. We may ask how they are contested. We may also ask how – and indeed whether – they might be done differently. In short, we open ourselves up to the possibilities of an ontological politics.

This is work to be done, though it is work that has to be done carefully.

First, it is important to understand that enacting realities is not a matter of volition. Whether or not a reality can be held steady in a practice – whether or not it will hold – is a practical matter. The ordering of practices turns around what one might think of as an intricate choreography of relations.[30] Think of a stakeholder meeting in these terms and the complexities implied in holding things steady start to emerge. Crucially, observe that intentions and designs – explicit designs – form only a small part of that choreography. Yes, PowerPoints and the programme of a meeting are designed. Again, it is not by chance that consumers are depicted as having particularly solid views on farm animal welfare. There is, in other words, an explicit politics of reality-making. But most of the relations assembled to do the meeting and its various realities were either designed elsewhere (think of the electricity supply, a crucial but unspoken component in the relations that made the meeting, or the computer software), or they happened anyway independently of intention. Think, for instance, of the bodies of the speakers; their clothes; the common language (English); the time coordination; the conventions (timeslots, talks, questions and answers, breaks and all the rest) within which the meeting was structured and ordered. Here is the point. All of these were a part of the ordering of the Berlin meeting. All participated in the realities enacted there. None could easily have been wished away. An attempt to do something different, very different, might have been possible but it would not have been trivial. Enacting realities is not a trivial matter.

So this is the issue. To wash away the metaphysics of common-sense realism is not to claim that anything goes. It is to shift our understanding of the *sources* of the relative immutability and obduracy of the world: to move these from 'reality itself' into the choreographies of practice. And then it is to attend to how the latter are done – and might be undone. But this shift also demands that we attend to the collateral realities – all those realities that get done along the way, unintentionally. For, here is my assumption: it is the endless enactment of collateral realities that tends to hold things steady. That (this is the tension) helps to make the choreography possible, but at the same time renders an ontological politics unthinkable. So what may be said of collateral realities?

Let me recap.

Put on one side, first, those realities that are being explicitly described or enacted: for instance, to do with the preferences of Hungarian consumers, or the need for animals to express 'intuitively desirable natural behaviours, such as exploration and play'. These are reals and they are being done, but they are being done in a manner that is articulated and made explicit. This means that it is easy to see them, and relatively easy to imagine that they might be done differently. Indeed they are contested. Attend, then, instead or in addition, to what is being done along the way, quietly and incidentally (See the figure below).[31] This we have seen already. But look at what it *does*. Here is a partial list.

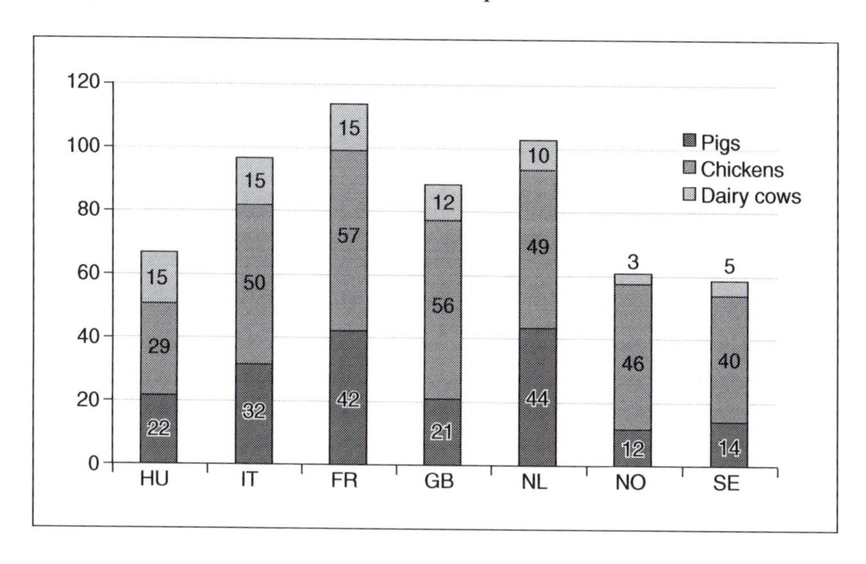

It does Hungary and six other countries. As a part of this (and as I have noted already) it does nation-states. It does publics as statistical aggregates of individual responses. More specifically, it does nation-states as statistical aggregates of individuals that may be distinguished from one another. It does surveys as an appropriate methodology for social research. It does the sample statistics built into surveys as an appropriate technique for deriving general claims about populations out of samples. It does questionnaires as workably reliable sources of data about people. As I have noted, it does individuals as quantifiable (actually self-quantifying) respondents. It does those individuals as containers of more or less stable attitudes that may be tested and determined in a questionnaire. It does something about transportability, by which I mean that it takes it for granted that questionnaire responses may be assembled, collated, summarised, but (this is the important point) moved from one appropriate location to another and still hold their validity and salience. It does something about the reliability and the relevance of social science survey research. As a part of this and the transportability, it does the possibility of a centralised viewpoint – so to speak the possibility and, no doubt, the need of a panoptical overview of country differences. And then, on top of all this, it does a series of metaphysical realities as well. So, for instance, if there is country geography, then so too there is something like Euclidean space. No doubt (as it happens this particular PowerPoint does not show this) if there are dates, then there is chronological time. It does a distinction between knower and known, subject and object. More specifically, it does a distinction between animal (object) and person (subject), and between human and non-human. And then, as a further part of this, it also does a distinction between the world on the one hand, and knowledge of the world (including statistics or numbers) on the other.

These are some, just some, of the collateral realities being done in this single PowerPoint. And most of them (this is why they are *collateral* realities) are being done incidentally and along the way, without any kind of fuss at all. As I have noted, one can imagine discussion about, say, the accuracy of the statistics about Hungary. Was the sampling appropriate? Or did the translation from English to Hungarian work? But nation-states? Statistical methods? Human and non-human? Reality, and knowledge of that reality? Or space and time? Probably, usually, these are realities that are not questioned. Rather, they work to *frame* what is being told more explicitly. But therefore, and very powerfully, this means that they are *also being done.* Here is the proposition: whatever is not contested and, more particularly, whatever lies *beyond the limits of contestability* is that which operates most powerfully to do the real. And it is this, to be sure, that is the technique which lies at the heart of common-sense realism. It is the enactment of collateral realities that turns what is being done in practice into what necessarily *has* to be.

Of course we cannot contest everything. Our own practices enact collateral realities like any others. We are no different. But this does not mean that we should not explore how practices do reals, and do so unintentionally and along the way. And indeed, as we have seen, there is something about the character of practices that will help us as we embark upon this adventure. This is the fact that they are never coherent. Earlier I wrote that in a series of PowerPoints, Blokhuis' talk was assembling a *succession* of Welfare Quality realities. These included the genealogy of a space-time box; a collaboration between animal scientists and social scientists across a scientific division of labour; and something that looked like a cybernetic system. What this tells us is that the reality that Blokhuis was assembling was non-coherent – not incoherent (this points us to a normative failure which is not what I intend) but *non*-coherent. Just, in fact, like the animal science, and the differences between surveys and focus groups.

Please understand that this is not a complaint about Blokhuis' talk, about animal science, about social research methods, or indeed about Welfare Quality®. Appreciate, instead, that it is an observation about the nature of practice. Coherence was the last of the features of common-sense realism that I listed. But coherence is simply an aspiration. In practice, practices are always more or less non-coherent. They work by enacting different versions of reality and more or less successfully holding these together. But if there is multiplicity rather than singularity then we have an entry point.[32] If we look for non-coherences within practices we will find them. We will discover collateral realities. And, this is the move to an ontological politics; we may take sides and hope to make a difference. Reality is no longer destiny.

Notes

1 I am grateful to Welfare Quality participants including especially Adrian Evans, Lindsay Matthews, Mara Miele and Isabelle Veissier for discussion and debate. Outside Welfare Quality, I am grateful to colleagues including Marianne Lien, Annemarie Mol, Ingunn Moser, Vicky Singleton and Wen-yuan Lin for their shared concern with the material–semiotic character of practice.
2 For further discussion, see Law (2004).
3 The position I am exploring has been developed within a particular version of STS (Science, Technology and Society) which includes actor-network theory and its successor projects, feminist material–semiotics, and versions of postcolonialism. For important writing in first of these see Callon (1998), Latour (1988), Law (2002), Mol (2002), Moser (2008), Singleton (1998) and Thompson (2002). For feminist material–semiotic work see Barad (2007) and Haraway (1997, 2007). For postcolonial writing see Verran (1998, 2001). For an introductory but specific survey of all three see Law (2008).
4 On empirical philosophy, see Mol (2002).
5 Welfare Quality (2007).
6 Welfare Quality (2008).
7 Blokhuis (2007a).
8 This comes from Blokhuis (2007b: 9).
9 Blokhuis (2007a).
10 This is Donna Haraway's metaphor See Haraway (1991).
11 For a related list, see Law (1986).
12 Blokhuis (2007a).
13 On the importance of 'projectness' see Law (2002: 183).
14 Blokhuis (2007a).
15 Blokhuis (2007a).
16 Blokhuis (2007a).
17 A related distinction that includes system and genealogy is developed in Law (2002).
18 For discussion, see Mol (2002).
19 See Law (1994).
20 There are huge literatures both on epistemology and on research methodologies (the latter especially in social science). The classic epistemological literatures offer two large narratives about method. The first is that the descriptions that it produces (should) correspond to reality. For a classic example see Nagel (1979). The second is pragmatic, and suggests that those descriptions are best understood as simplifying tools for handling a complex reality. For an example of the latter, see Kuhn (1970).
21 Blokhuis (2007b: 9)
22 Kjærnes *et al.* (2007).
23 For discussion see Waterton and Wynne (1999) and Law (2009b).

24 Law (2009a).
25 Veissier and Evans (2007a: 19).
26 Veissier and Evans (2007b).
27 Broom (1991: 4167).
28 Fraser (1993: 39).
29 Veissier and Evans (2007a: 19).
30 I borrow the metaphor from Charis Cussins (1998). See also Thompson (2002).
31 Kjærnes *et al.* (2007).
32 The point is explored at length in Mol (2002).

References

Barad, *Karen (2007), Meeting the Universe Halfway: Quantum Physics and the Entanglement of Matter and Meaning,* Durham, NC: Duke University Press.

Blokhuis, Harry (2007a), 'Welfare Quality®: Context, Progress and Aims', paper delivered at the Assuring Animal Welfare: from Societal Concerns to Implementation: Proceedings of the Second Welfare Quality® Stakeholder Conference, 3–4 May, Berlin, Germany. Also available at www.welfarequality.net/publicfiles/37404_657483129852_200705091222666_73238_Welfare_Quality_context_aims_and_progress_Harry_Blokhuis.pdf.

—— (2007b), 'Welfare Quality®: Context, Progress and Aims', pp. 9–12 in Isabelle Veissier, Björn Forkman and Bryan Jones (eds), *Assuring Animal Welfare: from Societal Concerns to Implementation: Proceedings of the Second Welfare Quality® Stakeholder Conference, 3–4 May 2007, Berlin, Germany,* Lelystad, the Netherlands: Welfare Quality®. Also available at www.welfarequality.net/publicfiles/36059_25646376170_200705090907523_2244_Proceedings_2nd_WQ_Stakeholder_conference_3_4_May_2007.pdf.

Broom, Donald M. (1991), 'Animal Welfare: Concepts and Measurement', *Journal of Animal Science,* 69: 4167–4175.

Callon, Michel (1998), 'An Essay on Framing and Overflowing: Economic Externalities Revisited by Sociology', pp. 244–269 in Michel Callon (ed.), *The Laws of the Markets,* Oxford and Keele: Blackwell and the Sociological Review.

Cussins, Charis M. (1998), 'Ontological Choreography: Agency for Women Patients in an Infertility Clinic', pp. 166–201 in Marc Berg and Annemarie Mol (eds), *Differences in Medicine: Unravelling Practices, Techniques and Bodies,* Durham, NC and London: Duke University Press.

Fraser, David (1993), 'Assessing Animal Well-Being: Common Sense, Uncommon Science', pp. 37–54 in Bill R. Baumgardt and H. Glenn Gray (eds), *Food Animal Well-Being,* Indiana: USDA and Purdue University Office of Agricultural Research Programs, also available at www.ansc.purdue.edu/wellbeing/FAWB1993/Fraser.pdf.

Haraway, Donna J. (1991), 'Situated Knowledges: The Science Question in Feminism and the Privilege of Partial Perspective', pp. 183–201 in Donna Haraway (ed.), *Simians, Cyborgs and Women: the Reinvention of Nature,* London: Free Association Books, also available at www.hsph.harvard.edu/rt21/concepts/HARAWAY.html.

—— (1997), Modest_Witness@Second_Millennium.Female_Man©_Meets_OncomouseTM: *Feminism and Technoscience,* New York and London: Routledge.

—— (2007), *When Species Meet,* Minneapolis and London: University of Minnesota Press.

Kjærnes, Unni, Emma Roe, and Bettina Bock (2007), 'Societal Concerns on Farm Animal Welfare', paper delivered at the Assuring Animal Welfare: From Societal Concerns to Implementation: Proceedings of the Second Welfare Quality® Stakeholder Conference, 3–4 May, Berlin, Germany. Also available at www.welfarequality.net/publicfiles/37404_657483129852_200705091238942_14184_Societal_concerns_on_farm_animal_welfare_Kjaernes_Roe_and_Bock.pdf.

Kuhn, Thomas S. (1970), *The Structure of Scientific Revolutions*, Chicago, IL: Chicago University Press.

Latour, Bruno (1988), *The Pasteurization of France*, Cambridge MA: Harvard University Press.

Law, John (1986), 'The Heterogeneity of Texts', pp. 67–83 in Michel Callon, John Law, and Arie Rip (eds), *Mapping the Dynamics of Science and Technology: Sociology of Science in the Real World*, London: Macmillan.

—— (1994), *Organizing Modernity*, Oxford: Blackwell.

—— (2002), *Aircraft Stories: Decentering the Object in Technoscience*, Durham, NC: Duke University Press.

—— (2004), *After Method: Mess in Social Science Research*, London: Routledge.

—— (2008), 'Actor–Network Theory and Material Semiotics', pp. 141–158 in Bryan S. Turner (ed.), *The New Blackwell Companion to Social Theory*, Oxford: Blackwell.

—— (2009a), 'Seeing Like a Survey', *Cultural Sociology*, 3(2): 239–256.

—— (2009b), 'Staging the Social: Notes on Surveys and Focus Groups', paper delivered at the Workshop: Publics: Embodied, Imagined, Performed, Cesagen, Lancaster University, 27–28 January.

Mol, Annemarie (2002), *The Body Multiple: Ontology in Medical Practice*, Durham, NC and London: Duke University Press.

Moser, Ingunn (2008), 'Making Alzheimer's Disease Matter: Enacting, Interfering and Doing Politics of Nature', *Geoforum*, 39: 98–110.

Nagel, Ernest (1979), *The Structure of Science: Problems in the Logic of Scientific Explanation*, London: Routledge & Kegan Paul.

Singleton, Vicky (1998), 'Stabilizing Instabilities: The Role of the Laboratory in the United Kingdom Cervical Screening Programme', pp. 86–104 in Marc Berg and Annemarie Mol (eds), *Differences in Medicine: Unravelling Practices, Techniques and Bodies*, Durham, NC: Duke University Press.

Thompson, Charis (2002), 'When Elephants Stand for Competing Models of Nature', pp. 166–190 in John Law and Annemarie Mol (eds), *Complexity in Science, Technology, and Medicine*, Durham, NC: Duke University Press.

Veissier, Isabelle, and Adrian Evans (2007a), 'Rationale Behind the Welfare Quality® Assessment of Animal Welfare', pp. 19–22 in Isabelle Veissier, Björn Forkman, and Bryan Jones (eds), *Assuring Animal Welfare: From Societal Concerns to Implementation: Proceedings of the Second Welfare Quality® Stakeholder Conference, 3–4 May 2007, Berlin, Germany*, Lelystad, the Netherlands: Welfare Quality®. Also available at www.welfarequality.net/publicfiles/36059_25646376170_200705090907523_2244_Proceedings_2nd_WQ_Stakeholder_conference_3_4_May_2007.pdf.

—— (2007b), 'Rationale Behind the Welfare Quality® Assessment of Animal Welfare', paper delivered at the Assuring Animal Welfare: from Societal Concerns to Implementation: Proceedings of the Second Welfare Quality® Stakeholder Conference, 3–4 May 2007, Berlin, Germany, Lelystad, the Netherlands, 3–4 May. Also available at www.welfarequality.net/publicfiles/36059

_25646376170_200705090907523_2244_Proceedings_2nd_WQ_Stakeholder_conference_3_4_May_2007.pdf.

Verran, Helen (1998), 'Re-Imagining Land Ownership in Australia', *Postcolonial Studies*, 1(2): 237–254.

—— (2001), *Science and an African Logic*, Chicago, IL, and London: Chicago University Press.

Waterton, Claire, and Brian Wynne (1999), 'Can Focus Groups Access Community Views?', pp. 127–143 in Rosaline Barbour and Jenny Kitzinger (eds), *Developing Focus Group Research: Politics, Theory and Practice*, London: Sage.

Welfare Quality (2007), 'Presentations and Workshops of the 2nd Welfare Quality Stakeholder Conference', Lelystad: Welfare Quality, www.welfarequality.net/everyone/37404 (accessed 8 January 2009).

—— (2008), 'Stakeholder Conferences', Lelystad: Welfare Quality, www.welfarequality.net/everyone/36059 (accessed 9 January 2009).

9 Transforming the intellectual[1]

Patrick Baert and Alan Shipman

The sociology of intellectual life is a relatively underdeveloped subfield of sociology. While sociological classics and key twentieth-century authors pioneered and developed the sociology of knowledge, few of their studies looked critically at the world of intellectuals and intellectual production. It is only in the course of the past two decades that the subfield of the sociology of intellectual life has become more prominent, mainly due to three important developments: Pierre Bourdieu's reflexive sociology (Bourdieu 1988, 1991, 1996, 2000), Randall Collins' network approach (Collins 1998) and Charles Camic's new sociology of ideas (Camic 1983, 1987). From different perspectives, all three developed theoretically or methodologically sophisticated attempts to study intellectuals, in particular philosophers. What is striking, however, is that these approaches tend not to address the phenomenon of *public* intellectuals or, more broadly, how intellectuals engage with communities outside the narrow confines of the academy. On the contrary, they very much explore social mechanisms within academic institutions and within intellectual circles. There is little interest in how ideas, held by a small circle of specialists or experts, eventually reach a broader audience, or how interaction with that audience might affect intellectual projects and exchanges.

An intellectual is defined (in these authors' work, and in this chapter) as an individual who creates knowledge about an aspect of the natural or social world, and relates this to existing knowledge. Knowledge is defined in its broadest sense, as communicable ideas that convey cognitive value (Young 2001, Rouanet 2003). It therefore encompasses the artistic and religious as well as the scientific, and reasoned opinion as well as demonstrated 'fact'. The intellectual pursuit of knowledge goes beyond its practical pursuit for the fulfilment of specific tasks, as performed by researchers in universities or industry and 'knowledge workers' employed in the private or public sectors. While intellectuals may conduct research on closely defined topics and focus on solving particular problems, their role consciously extends to contextualising new knowledge, linking new ideas to those that already exist, building theories that account for a broad range of observations or empirical discoveries, and refining or extending the application of existing theories.

A public intellectual is a knowledge creator who addresses issues of social concern, and engages with a broader public. Whereas intellectuals value knowledge in itself and develop it through interaction with other intellectuals, a public intellectual identifies additional value in publicly presenting and applying knowledge, and so addresses an audience beyond intellectual peers. This engagement can take place in different ways. The 'media-enhanced' public intellectual gains access to the public with the help of journalism, broadcasting, book and journal publications and new media. They tend to acquire a strong public profile. The 'direct' public intellectual gains more immediate access to the public: through (for example) delivering public lectures, being called as an expert witness at a trial or public enquiry, or advising a social or political organisation. Direct engagement may not generate a public profile, but increasingly has the potential to do so as new communication media (including print and audiovisual copy recorded by the public and transmitted via the internet) bring the initially localised message to a wider public.

When northern hemisphere sociologists and historians have written about public intellectuals, they have mainly done so in a restricted fashion, promoting what we call a 'declinist' argument. Declinists point to a previous golden age of the public intellectual and argue that their number and influence has been falling for some time (Jacoby 1987; Posner 2001; Furedi 2004). Their downbeat assessment is complemented by studies of the 'knowledge economy' which depict a growing proportion of society applying knowledge at work, but few developing their ideas intellectually – and many who might have been intellectuals being drawn into commercialised, career-building alternatives (Florida 2002: ch. 11; Brooks 2001). Although the advance of knowledge may lead to a larger 'community' of intellectuals, declinists suggest that a greater proportion of them have become privatised, not expecting people outside that community to understand their work or shape its application. As well as fewer intellectuals becoming public, declinists also identify a loss of public intellectuals' power and influence in society. There is often a clear moral dimension to this argument: public intellectuals are portrayed as providing a positive impact, contributing to a strong civil society and healthy democracy, and their decline is seen as inviting a bleaker future in which policy and institutions are less amenable or accountable to the public. However, many declinists also present the decline of the public intellectual as the obverse of non-public intellectuals' rising numbers and influence within the intellectual community. Those who stay focused on an intellectual audience contribute to undermining the authority of those who 'go public', by characterising their broader perspective as populist and superficial. Decline becomes self-reinforcing once public intellect is viewed as oxymoronic, with publicity compromising intellectuality.

This chapter argues that, even in the Anglo-American context in which it has principally been argued, the decline or disappearance of the public intellectual is largely a misperception. The declinist argument tends to understate the extent to which intellectuals can (and do) still broaden their motivation and social engagement to become public intellectuals; and it tends to overstate the extent to which, conversely, intellectuals can (and do) narrow their motivation and social engagement to become 'experts' or 'technocrats'. We go on to argue that the range of public engagement strategies used by intellectuals has actually widened over the past century, as a result of changes in the activities of intellectuals, their institutional context, and the nature of the public with whom they potentially engage. Although declinists are correct in identifying a large cohort of intellectuals who are no longer publicly engaged, they are wrong to imply that the number or influence of public intellectuals has declined. This is, in part, because they fail to recognise new *types* of public intellectual, made possible (and in some cases necessary) by institutional and cultural changes, and new *sources* of public intellectual, especially from the natural sciences. These changes have arisen from the impact of intellectual activity on knowledge and perceptions of knowledge, as well as from changed external demands on the process and products of intellectual work. So the chapter ends by identifying the new routes through which public intellectuals emerge from within larger intellectual communities.

The 'declinist' thesis

Two proximate causes of public intellectual decline have been suggested. The first, intra-institutional type of argument is epitomised by Russell Jacoby's position and puts the blame squarely on modern universities (Jacoby 1987), for absorbing intellectuals and restricting their capacity for critical or innovative thinking. It corresponds to the romantic image of the intellectual who is best able to engage with a broader public when not institutionally curtailed. The employment and career structure within university institutions is held to be incompatible with the role and ethos of the public intellectual. The specialist subdivision of disciplines and the rise of the bureaucratically managed 'multiversity' (Kerr 1963) pushes academics into disciplinary silos without the incentive to engage with the outside world, or even with other disciplines. Intellectuals had a distinctly public profile until they were employed *en masse* within universities. Ironically, while the expansion of the university sector in the 1960s was accompanied by a political radicalisation of the student movement, the very same increase was also a contributory factor to the erosion of the public intellectual. University 'domestication' implies that, for intellectuals to obtain and retain a position and to be promoted, they have to address other intellectuals rather than a broader audience and develop a technical, jargon-laden language, addressing increasingly esoteric issues. Peer approbation eclipses societal significance as a criterion for intellectual success. The importance of clustering, for the creation and propagation of critical ideas, means that economic geography has also played a role in decline. In the early twentieth century, intellectuals' survival without secure employment was assisted by low rents. Once rents increased, many were forced to take university positions which often meant moving away from urban centres, to a suburban life (and commuter travel) less conducive to political or social engagement.

The second, extra-institutional type of argument, though often compatible with the first, focuses on structural changes around universities, especially in their relation to the economy and the state. Whereas the previous argument emphasised the isolated, self-referential nature of modern universities, here the argument is that universities are too much intertwined with outside economic interests that discourage critical thought. For example, Nisbet ([1970]1997) argues that after eight centuries of preserving free thought insulated from worldly pressures, universities in the twentieth century succumbed to annexation by governments that saw them as a means to boost national income and employment, and businesses as a source of practical organisational and technical knowledge. Their transition from an enclosed mediaeval institution to one exposed to modern commercial pressures is viewed as having destroyed the pursuit of learning for its own sake. Intellectuals are now employed to serve society (by raising its productivity and solving its operational problems), and so are deterred from taking on a more public role that might question the workings of society and argue for alternatives. Evans (2005) suggests that the corruption of universities' knowledge-creating mission has been matched by an instrumentalising of their teaching, which is now more aimed at cramming students for exam success than encouraging them to challenge accepted ideas and think for themselves.

The intra- and extra-institutional arguments offer different explanations for the same phenomenon: a diminishing proportion of intellectuals becoming public intellectuals, and an increasing proportion of intellectuals migrating to a more restricted role: that of the 'expert' or 'technocrat'. This shift is underlined by changes in the way intellectual status is defined and assigned. When the term was first deployed, most intellectuals were 'public', deploying their skills to pursue a social or political cause (Eyerman 1994). In the century since 'intellectual' was first deployed in this way, it has also come to be applied to a new type of professional who pursues the creation and dissemination of knowledge without directly engaging with the public, usually in the setting of a university. Those going 'public' are now a subset of the whole.

The specialisation of avenues and audiences for intellectual enquiry gives rise to a new social role, that of intellectual expert or technocrat. These are individuals who receive an academic or professional training, via a university or equivalent institution, but who restrict its use to narrowly contextualised problem-solving, even if principally employed in an academic post. In contrast to intellectuals, experts are viewed as accepting and implementing received ideas and procedures, seeking to vindicate them by results rather than critically examining their principles. Commentators have coined different labels to refer to the rise of the experts. For example, some characterise this in terms of a shift from 'double-' to 'single-loop' learning (Argyris 1976) or from the theoretically grounded 'Mode 1' to the application-oriented 'Mode 2' variety of knowledge (Gibbons *et al.* 1994). The common theme is that experts are viewed as applying ideas in pursuit of results, not developing ideas for their intrinsic merits or contextualising and connecting them.

The 'non-intellectual' expert tended, initially, to be identified especially with the natural sciences. Physics, chemistry, engineering, mathematics and computing were acknowledged to have helped the Second World War effort and peacetime economic recovery, but those who advanced and applied them were widely viewed as lacking the breadth of interests and contextual understanding of their work that would qualify them as intellectuals. Even one of the staunchest admirers of post-war scientists was forced to concede that "the whole literature of the traditional culture doesn't seem to them relevant to those interests.... As a result, their imaginative understanding is less than it should be. They are self-impoverished" (Snow 1959: 13–14). However, this criticism has subsequently been extended to intellectuals in the social sciences, arts and humanities. Increased specialisation, leading to greatly increased complexity and quantity of analysis in every field, makes it mentally harder to maintain broader interest and understanding.

Motivationally, the need to forge a career within a state- or corporately financed university makes it materially less rewarding to do so. Intellectuals are seen as seeking routine applications of established knowledge, to sustain and strengthen the corporate economy (and tax base) that provide their career structure – hence rendered unable or unwilling to mount disruptive challenges to that knowledge which might shake up their institutional surroundings or force them to think afresh. So, for example, Boggs (1993), building on earlier observations of Gramsci, argues that economic progress (in Western liberal democracies) is associated with the eclipse of 'critical' intellectuals (who challenge accepted wisdom, maintaining a pluralism of methodologies and theories) by 'technocratic' intellectuals, who narrow the task down to one of problem-solving with restricted, even formulaic consensus procedures. From disparate political backgrounds, these thinkers identify a tendency for intellectuals to cease constructively challenging, and start conforming to, a system that harnesses knowledge for wealth creation and gives up the autonomy of knowledge creation. Some Eastern European and North American sociologists have gone even further and identified intellectuals as a 'new class' that exercises power through control of intellectual rather than material property, but which substitutes conformist for critical thinking in the process of taking that power (Djilas 1957; Konrad and Szelenyi 1979). This criticism initially from communist-era Eastern Europe was echoed by North American liberals at the height of post-war capitalist recovery. For them, "the technostructure has become deeply dependent on the educational and scientific estate for its supply of trained manpower ... [and] what may be called the reputable social science no longer has overtones of revolution. Rather it denies the likelihood, even the possibility" (Galbraith 1967: 290).

Although Nisbet, Evans and Boggs do not explicitly address the issue of public intellectuals, the negative implication is clear: once universities are compromised and thinkers and scholars no longer take seriously their critical function and their role in engaging others to think critically, the role of public intellectual is under threat. If bureaucratically restructured, economically captured universities stifle critical thought, then any surviving public intellectuals are likely to have moved away from academia, or never been immersed in it. Before Western universities' expansion in size and number accelerated, many public intellectuals had operated outside them; and the bureaucratic constraints that accompanied this expansion can still be an inducement to remain outside, even when this means more distance from collaborators and sources of funding. For example, James Lovelock has promoted the 'Gaia Hypothesis' as an independent scholar, consulting for academic and commercial organisations but avoiding employment by them, and making supplementary income through popular book sales.

> Science was and is my passion and I wanted to be free to do it unfettered by direction from anyone, not even by the mild constraints of a university department or an institute of science. Any artist or novelist would understand – some of us do not produce their (*sic*) best when directed.
>
> (Lovelock 2000: 2)

Likewise, the journalist Christopher Hitchens, reflecting on his nomination as a leading public intellectual, argues that academic status is now a barrier to engagement with the public or to the production of radical ideas with which the public would want to be engaged. For him, the term 'public intellectual' "largely describes people who worked outside the academy and indeed outside of large-scale publishing, tending to be self-starting independents or editors of 'minority-of-one' type magazines" (Hitchens 2008).

The 'declinist' argument also underpins Ben Agger and Michel Burawoy's argument that sociology ought to go beyond the safe contours of the academy and take on a public, political role (Agger 2000; Burawoy 2005). Their argument is that sociology fought hard to obtain professional and academic status and broader recognition, and that in this process it has lost some of its previous public functions. In so transforming, sociology follows a path repeated by other intellectual disciplines as their practitioners expand in number and move inside the academy. They form an increasingly enclosed and isolated community that – by virtue of concentration, separation from society, and specialisation of language and models – puts dialogue within itself above engagement with the wider community. They also, through peer review, become the sole arbiters of valid and applicable knowledge, reinforcing a vertical, one-way knowledge transmission role towards students and the rest of society. Now that sociology has obtained this long-sought recognition, it needs to shed this defensive shield and strive for a public profile, assisting and directing various political and social movements. While Burawoy's plea for a public sociology is an attempt to revive what he sees as a previous golden era of publicly orientated sociology, he is also anxious to demarcate his project from it. Whereas early twentieth-century sociology was what he calls 'traditional' public sociology, he is arguing for 'organic' public sociology. Whereas the former addresses an amorphous, passive, invisible audience and lacks a clear social or political agenda, the latter targets a visible and active counter-public with an explicit agenda.

In calling for a revival of public sociology, Burawoy asserts that the discipline has, in the course of the twentieth century, retreated from the public sphere and ceased to produce public intellectuals comparable to W.E.B. Du Bois, David Riesman or Robert Bellah. He contrasts a romantic image of a golden era of public intellectual life with the barrenness of the current age. Like Jacoby, Burawoy suggests that professionalisation and academic institutionalisation have made it increasingly difficult for public intellectuals to emerge: the self-referential nature of academic life subverts the outward nature of the public intellectual sphere. Academics can now address their work entirely to other academics, who are the only audience that count for validation through peer review. Their work thus becomes an internal conversation; far from seeking to address the public, they consciously exclude it by confining their debates within technical terms and academic journals.

The declinist backdrop: wider intellectual decline

As well as these explicit commentaries on the decline of public intellectuals, a number of recent studies convey a complementary implicit argument: that intellectuals *in general* have lost social status and influence, and that intellectuals seek to regain these by following strategies that move them away from any public role. Three types of explanation, epitomised by Zygmunt Bauman, A.H. Halsey and Karl Popper, are frequently advanced to account for the declining social importance of intellectuals, and the consequently reduced ability or willingness of intellectuals to engage with wider publics.

The first line of argument is that the notion of the intellectual as 'legislator' is in decline (Bauman 1991). The Enlightenment vision assumed that intellectuals obtain superior knowledge and ethical and aesthetic judgement by transcending culture and language. Bauman claims that this view is no longer prevalent today owing to widespread scepticism towards foundationalist philosophy. Instead, the postmodern condition promotes a view of intellectuals as mere 'interpreters', skilful in making sense of one culture to another. Although Bauman does not explicitly address the issue of public intellectuals, it is clear that the Enlightenment notion of the intellectual as legislator is conducive to their emergence. Indeed, foundationalism provides the intellectual with the necessary back-up and authority to enter the public domain. Therefore, evidence of the erosion of the intellectual as legislator could be taken as evidence for the decline of the public intellectual.

The second approach is exemplified by Halsey's 'the decline of donnish dominion' (Halsey 1992), which identifies social causes of the erosion of intellectual authority. Halsey argues that the twentieth-century expansion led to a 'democratisation' of knowledge production and consumption which deprived the dons of their former dominion over society. The vast expansion in the number and size of universities, moves from small-group to classroom tuition, and the arrival of new (often modularised) degree disciplines eroded academics' status, while multiplication of the number of graduates narrowed the gap between 'intellectuals' and the wider public. While ancient universities (in England, Oxford and Cambridge) retained their prestige, they became small islands in a sea of newer institutions whose academics found less research time amid teaching duties, felt less valued, encountered more external pressure over research and curriculum design, and worked in discipline-divided faculties rather than broader collegiate settings. Publicly funded expansion also makes universities vulnerable to a commercialisation and bureaucratisation that undermines their pursuit of unbiased knowledge for its own sake. In opening up higher education to larger segments of the public, governments have also exposed it to political pressures to pursue knowledge that is economically useful, and market pressures to give students what they need to pass exams and gain good career tracks. Like Bauman, Halsey does not explicitly discuss the notion of the public intellectual. His surveys suggest a general trend (especially in more recently chartered universities) for wider public engagement, in the sense of accepting students from more varies socio-economic backgrounds and dealing with more outside agencies in procuring funds for research or conveying its results. But this is not intellectual engagement. The structural factors which Halsey identifies as eroding the intellectual authority of academics inevitably deprive them of the authoritative voice to become effective public intellectuals.

A third type of argument that may underscore the declinist argument arises from the philosophy and social studies of science. It suggests that changes in the way knowledge is generated and validated (in the natural and social sciences) have lessened the impact of intellectual work on public opinion, and made it less likely that intellectuals will gain public prominence. The philosophy of science, in moving against positivism and the deductive-nomological method, has created an impression of scientists' discoveries and opinions as more provisional, fallible and limited in scope than they (and the public) once believed (Popper [1963]2002). Specialisation has undermined past holistic and 'grand' theories without creating new ones, and the multiplication of scientific effort has tended to prevent any unanimity, with conflicting results or dissenting interpretations always available. Where a consensus is reached, its spreading across many equally qualified professional scientists may reduce the likelihood that one will be viewed as the pioneer or champion and emerge into public view. So, paradoxically, the 'growth of knowledge', the increasing sophistication of scientific method and the rising number of practising scientists seems to be accompanied by a decline in the status and power of the intellectual. Knowledge and its production become subjected to democratic pressures, with intellectuals inside and outside universities competing with other social actors for its construction and validation (Delanty 2001; Friese and Wagner 1998).

Contesting decline: the renewed intellectual engagement

The argument that public intellectuals have fallen in number and influence, in a context of general intellectual retreat, is deficient in at least two ways. First, authors who subscribe to the declinist argument tend to distort the past and present: they romanticise and glamorise intellectual life in the early twentieth century while depicting the present as devoid of public intellectual life (Etzioni and Bowditch 2006). American intellectuals are particularly keen to depict their own intellectual history in this light, thereby attempting to argue that we need to revive what they see as a truly American pragmatist tradition that conceives of intellectual life in terms of social and political engagement. This leads them to understate the degree to which intellectuals could remain publicly engaged even after recruitment to universities and absorption into academics' internal conversation. A substantial amount of American public intellectuals during the so-called golden era of the public intellectual, like John Dewey or David Riesman, held academic positions. The student uprisings of the 1960s drew intellectual and practical inspiration from academics (notably Herbert Marcuse), and the explosive growth of interest in communication and new media was similarly sparked and steered from within universities (by such figures as Marshall McLuhan, Neil Postman and Raymond Williams). The past two decades have not seen the decline of public intellectual life. On the contrary, there have been a significant number of writers with a strong public profile, including many who hold or have held academic positions: Judith Butler, Noam Chomsky, Anthony Giddens, John Gray, Germaine Greer, Paul Krugman, Edward Said and Joseph Stiglitz are among many who have substantially influenced public debates and attained associated public profiles without any compromise to their intellectual status.

Second, authors who advocate the declinist thesis tend to present public intellectual life as an amorphous entity: they make little distinction between different types of intellectual life and different types of public intellectuals. Besides Burawoy's distinction between traditional and organic public intellectuals, there is little conceptualisation of the different shapes which public intellectual life can take. Consequently, they tend to equate subtle changes to public intellectual life with its decline or disappearance. The decline of early twentieth-century forms of intellectual authority, based on epistemic certainty and institutional prestige, leads not to the demise of public intellectuals but to the possibility of new types. Today's public intellectuals no longer rely on earlier types of legitimacy, but thrive on the prevailing sense of epistemic uncertainty and draw on an educated public to rally support for their views. Declinists fail to observe significant changes that are going on in the field of knowledge production and public intellectual life.

A distinction may be drawn between three ideal-types of public intellectuals: those whom we call authoritative intellectuals, professional intellectuals and embedded intellectuals. Although these three types can exist simultaneously, their relative numbers and significance have changed over time. Whereas authoritative public intellectuals tended to be characteristic of the nineteenth to early twentieth century, professional intellectuals emerge in the course of the twentieth century, and the past three decades have seen the rise of the embedded intellectual. These are acknowledged from the start to be ideal-types. In reality most public intellectuals combine features of all three; embedded intellectuals have existed in the past, just as authoritative and professional intellectuals still operate today. Nevertheless, there is a discernible trend away from the model of the authoritative figure to the professional intellectual and then to the embedded intellectual. Those who support the declinist thesis erroneously take the fall of the authoritative intellectual figure as a decline of the public intellectual as such. In this section we define our categorisation and explain why, although the number and proportion of intellectuals in each may have changed over time, nothing in such changes has undermined the relevance or prevalence of the public intellectual.

Authoritative intellectuals

The derivation of intellectual status from *authority* is common when professionalisation of intellectual life and specialisation among disciplines are at a relatively early stage, and when only a small proportion of the public receives higher education or a schooling that could lead to it. While modern universities established themselves in the course of the nineteenth and early twentieth century, this period pre-dates the intense professionalisation of academic disciplines that characterises the mid- to late twentieth century. This means, first, that universities have not yet acquired the monopoly of knowledge creation and dissemination. Scholars outside the university have a significant impact on intellectual life, and academics have to engage with them. For example, academic biologists had to engage with Darwin and amateur archaeologists and naturalists; and academic historians with Thomas Carlyle and Lord Acton (for the forty years before his first and only academic post). Second, the demarcation between academic disciplines is not fully developed and scholars are able to negotiate different fields of inquiry without losing credibility: for instance, Thorstein Veblen and Max Weber published both as economists and sociologists, Lord Kelvin combined mathematics, physics and engineering, and John Dewey's work was central to areas of philosophy, education and psychology. Third, there is limited standardisation within the disciplines in that there is not much of a consensus yet about the types of questions that ought to be asked and how to answer them. This means that people within a discipline are not able to present themselves in a unified fashion. For economics, until the 1920s when Marshall became influential, there was not a single agreed set of principles that made up the discipline and allowed it to be taught as a stand-alone subject (Kadish 1993). Those who establish such principles and assemble a curriculum around them can exert substantial power over the direction of thought in newly emerging disciplines, as Marshall demonstrated with his *Principles of Economics* (Marshall 1920) and launch of a degree in the subject at an ancient university.

Authoritative intellectuals have distinct charismatic qualities which make them particularly well suited for a public profile: they know how to write or speak with clarity and conviction, how to shock and how to entice their audience. They often help to fashion new disciplines, persuading the public that a subject area exists and is worth pursuing, as well as expressing what they claim to be accurate knowledge within that area. Although authoritative intellectuals may have a present or past university association, they do not derive their authority from being steeped in an academic discipline or from being an expert. They tend to be generalists who write and speak about a wide range of topics. The writings and public performances of authoritative intellectuals have a strong moral component: they exhibit character, take the moral high ground and thrive on adopting an outsider position. They rely on what we call 'vertical authority' in that they speak at, rather than with, their audience, even if they are populists or aim to promote equality. The dilemma of authoritative intellectuals, as it manifests itself in the course of the twentieth century, is intertwined with the process of professionalisation. As intellectual life increasingly becomes centred around what is called 'organised knowledge' and disciplinary specialisation, it becomes more and more difficult to act as generalists without being dismissed as dilettantes.

Professional intellectuals

Partly due to the success of authoritative intellectuals, the first part of the twentieth century saw an intensification of academic professionalisation. The rise of the *professional intellectual* is characterised by four developments. First, the major universities became dominant research institutions, with knowledge creation, dissemination and validation taking place increasingly within the academy. Universities also became centres of knowledge validation by collecting together intellectuals who discussed and assessed one another's work. Second, peer recognition became central to the dissipation of ideas. Ideas were increasingly only classified as 'knowledge' and legitimate if they appeared in peer-reviewed journals or were directly conveyed by researchers whose work had been approved by other experts. Intellectuals' power to pronounce on appropriate policies and public choices derives from the evidence and analysis they present rather than the status they hold. Third, standardisation took place as academics defined the boundaries of their disciplines and reached a relative consensus as to which language, methods and techniques to adopt for which purposes. Fourth, because of the standardisation, scholars were able to present themselves in a unified fashion and distinguish themselves from lay people. Scholars became increasingly preoccupied with refining methods and techniques that distinguish professional research from lay activities. They created academic languages and dialogues that distinguish intellectual knowledge from 'ordinary language' discourse and 'lay' dialogue.

In general, the humanities and social sciences have been slower than the natural sciences in adopting this professional model. With regard to standardisation, for example, there are clear differences within the humanities and social sciences in the US. On one end of the spectrum, sociology has remained a relatively heterogeneous discipline with different views as to what to study and how to study it, despite growing agreement as to which theories and methods are appropriate for which types of questions and even more consensus as to which journals are significant. In contrast, economics achieved an early uniformity. Economics has for half a century shown a relatively high degree of consensus owing to the dominance of rational choice theory and mathematical modelling based on constrained maximisation. Heterodox economists are unlikely to find a home in economics departments of major research universities in the US, and more likely to obtain a position in management schools which house different disciplines and preserve heterogeneity. Compared to sociology and economics, philosophy occupies a middle ground. A certain level of standardisation took place after 1945 when analytical philosophy managed to seize a dominant position in a number of key departments in the US; other forms of philosophy tending to be reclassified as 'history of philosophy' or exiled to other departments. However, philosophy at Yale kept its pluralism, and even at Harvard, the history of philosophy remained important throughout the 1960s. Outside the top American departments (and indeed outside the US), philosophy remained an ecumenical field.

Although running at different speeds in different disciplines and countries, these developments made possible the rise of the professional intellectual, who is steeped in a particular discipline, and whose authority draws on a considerable amount of expertise, recognition from peers and institutional back-up. This type of intellectual makes statements that draw on epistemic or logical certainty, referring to corroborated 'findings' or 'reasoning'. In its archetypal forms his or her public performance is in tune with the body of knowledge that is accumulated and which the public intellectual represents. This type of intellectual faces a 'differentiator's dilemma': epistemic or logical certainty and methodological consensus provides their legitimacy, but also implies standardisation which is not in tune with the charismatic qualities associated with public performance. When professional expertise opens up a role in consulting, advising and problem-solving in areas of public or commercial policy, they may also confront an 'innovator's dilemma': new discoveries which radically alter the discipline's core beliefs or procedures would render some of the expert knowledge redundant, so there may be an incentive to accommodate anomalous findings within the current framework rather than break out of it, as earlier intellectuals might have aspired to do.

'Declinists' have often treated professionalisation as tantamount to the disappearance of the public intellectual, and the downgrading of intellectuals in general, because it turns each discipline into a standard set of techniques and results which (in the manner of Weberian bureaucratisation) can be detached from their intellectual creators, to be learnt and applied by competent non-intellectual practitioners. In practice, there are two ways in which intellectuals can overcome this dilemma and still attain public prominence after their discipline becomes professionalised. They can excel at distilling and popularising a rising or dominant view within their discipline (for example, Richard Dawkins' best-selling presentation of *The Selfish Gene* (Dawkins [1976] 1989) develops and conveys an approach to evolution formulated in mathematical obscurity by William Hamilton (1964) and others, and Paul Samuelson wrote the standard postwar textbook on what he presented as 'Keynesian' economics) (Samuelson 1948). Or they can command attention by attacking an established view and developing an alternative, enhancing their public profile through the image of battling to overthrow an entrenched 'mainstream'. So, for example, Albert Einstein, Werner Heisenberg and Ilya Prigogine all presented radical alternatives to existing views of the natural world, Noam Chomsky's universal grammar (Chomsky 1957, 1965) counterposed earlier behaviourist approaches to linguistics, and Kenneth Tynan became a celebrated critic by propounding the merits of new forms of stage and television drama (Shellard 2003).

Some intellectuals span both groups by using a popularising agenda to pursue an iconoclasm. Textbooks and media appearances that present them to the public as representatives of the mainstream may be used to promote ideas which are displaced from, or subversive of, the mainstream. This may enable their alternative ideas to move towards mainstream acceptance faster than if they had not accessed a public channel through popularisation. Thus some evolutionary theorists accuse Dawkins of advancing the 'selfish gene' as a consensus view, marginalising rival theories that allow for group selection. Some economists complained that Samuelson popularised a 'Keynesianism' that lacked many of Keynes' more radical insights, keeping 'post-Keynesian' theories off the post-war curriculum; and a generation later, popular Anglo-American understanding of economics was reshaped by television documentaries from Milton Friedman (1980) and John Kenneth Galbraith (1977), both well outside the mainstream of the time. Some cosmologists worry that Stephen Hawking's views on space and time are now regarded as representative of the discipline, because of the millions who have bought his *Brief History of Time* (Hawking 1988) or viewed related documentaries, even though rival views may have comparable status within the cosmological community.

Those who obtain prominence in either of these ways can often expand it by engaging with the public in areas outside their discipline, drawing on their legitimacy in one area to exert influence in others. High status in a discipline regarded as rigorous, with a high degree of consensus over major discoveries, can be especially effective in enabling public pronouncements on political or social matters, where less consensus reigns. Thus, for example, nuclear physicists Robert Oppenheimer and Andrei Sakharov made notable attacks on their respective political systems in the 1960s, their scientific eminence offering comparative (but not complete) protection against reprisal. Noam Chomsky has used intellectual prominence in linguistics as the foundation for similarly prominent dissent over political issues (e.g. Chomsky 1983, 2003), as did the historian E.P. Thompson, whose popular critiques of defence policy (e.g. Thompson 1985) derived considerable status from his earlier intellectual work (e.g. Thompson 1963).

The notion of the professional intellectual has increasingly been undermined by a number of factors. First, there is growing recognition, within and beyond academic circles, that scholars' claims are rarely backed by epistemic certainty or logical proof. Philosophically, anti-foundationalist philosophies, like neo-pragmatism, have taken centre stage and question previous theories of truth and correspondence (e.g. Rorty 1979, 1991: 63–77, 1998: 19–42). The wider public is increasingly aware that there are limits to human knowledge, including limits to the knowledge of scientific experts (Smithson 1989; Gross 2007; Beck 2009: 115–128). For example, public policy has had to recognise that some decisions previously based on probabilistically analysed 'risk' actually involve genuine uncertainty, where we cannot attribute probabilities to possible outcomes (Callon *et al.* 2009: 13–36). The limitations of scientific knowledge are illustrated by recent fierce and unresolved debates over the dangers of volcanic ash for flying over Northern Europe, the causes of the 2007 to 2009 financial crisis and economists' failure to predict it, and with the possible negative side-effects of low frequency electromagnetic fields. In neither case is more research likely to do away with the genuine uncertainties that are at stake.

Second, the fact that more people are educated at high level makes it more difficult for intellectuals to sustain the type of vertical authority that is associated with the first two types of public intellectuals. With education comes a growing awareness of the fallibility and contested nature of scientific findings or philosophical positions, and indeed a willingness to challenge those views and to rally support to mount such a challenge. The intense public debates and parliamentary inquiries into academic conduct at the University of East Anglia's Climate Change Unit, after a leaking of emails (Marshall 2009; Corbyn 2010; Harrabin 2010), illustrate the greater openness of intellectuals to public scrutiny and criticism when pronouncing on issues with wider social and political impact. The increased use of academics as expert witnesses in criminal trials has exposed their reasoning to critical scrutiny, sometimes leading to arguments being undermined in the courtroom or by subsequent reanalysis. For example, after the testimony of eminent paediatrician Roy Meadow led directly to a mother's imprisonment for the deaths of her children, re-examination of his statistical methods suggested that he had substantially overstated the odds against their dying of natural causes, leading to the woman's release and the expert's disgrace (Dyer 2005). Increasingly, laypeople feel entitled to be involved in debates of this nature. Although few are equipped to read the scientific papers and question the conclusions directly, they have found ways to force intellectuals from 'lecture mode' into dialogue: probing and publicising the sources of evidence, deploying dissenting intellectuals who can move the peer-review process into public forums, and forcing intellectuals to state their case in 'ordinary language', which sometimes leads them to use metaphors and examples that are more easily attacked than the underlying model. The more 'vertical' an intellectual's pronouncements become, the stronger are they assailed by counter-forces aimed at knocking them down from their high perch, if their pronouncements significantly impact upon other groups or individuals.

Third, whereas previously the public financing of universities was relatively stable, increasingly universities are competing for funding, not just in the natural sciences but in the social sciences and even in the humanities. But what is worthy of funding is a political issue, and academics are increasingly embroiled in battles of persuasion over what is in society's interest to be funded. Even where they do not regard public opinion as having any direct relevance to knowledge creation, intellectuals are forced to recognise the indirect influence of the public in determining which research and teaching activities are worth supporting, and in influencing political decisions on the allocation of resources. Academics are becoming entrepreneurs whose lifeline depends not only on the approval of peers, but on the backing of administrative agencies whose priorities are shaped by the wider public. Public engagement has become vital to securing the inputs to knowledge creation, and ensuring a continued public appetite for the outputs.

Embedded intellectuals

Once a critical public needs to be convinced that intellectual activity has inherent merits and worthwhile products, the *embedded intellectual* becomes more prominent. The concept of embeddedness conveys a sharp reduction in, or disappearance of, intellectuals' ability to exist apart from the rest of society – in the double sense of maintaining a conversation, and resourcing that conversation, without engaging those outside the intellectual community. Whereas authoritative intellectuals could regard themselves (and be regarded by others) as 'above' society because of superior general knowledge and insight, and professional intellectuals could be similarly regarded as 'outside' society because of superior specific knowledge and expertise, 'embedded' intellectuals are inescapably immersed in their society. Maintaining their conversation requires them increasingly to engage in dialogue with others, and/or to persuade others to resource their efforts and endorse their results. Although intellectuals still change register when switching from an audience of peers to one of the public, fewer are happy with the idea that a gulf between 'lay' and 'learned' language reflects a gap in understanding or a need for conceptual dilution. More are likely to side with Amitai Etzioni in arguing that when published chapters "are made in two voices: that of the public intellectual and that of the academic … While the voice changes as different audiences are addressed, the points do not" (Etzioni 2001: xiv).

Embedded intellectuals manifest themselves in two different ways. One group, who might be termed 'intellectual persuaders', use an initially established role within an intellectual profession to build public and political opinion in favour of further institutional support (especially research funding), which then enhances the research programme. For example, Linus Pauling used his eminence in biochemistry (culminating in a Nobel prize) to mobilise support for his controversial views on 'orthomolecular' medicine, using public opinion to raise the profile and financial support for research linking diet to cancer; and Rupert Sheldrake has enrolled the public in experiments to test his highly controversial 'morphic resonance' theories (Sheldrake 1987). Increasingly, social and natural scientists draw on the media to drum up support for their research and their views. In the arts, intellectuals actively engaged in literature, drama and visual art have led campaigns for state sponsorship to replace past philanthropic sponsorship, and those in critical roles have appealed to the public as well as to policy-makers in establishing these as academic disciplines. They are not public intellectuals in the traditional sense of the word but they do engage with the public and the promoting exercise is central to the survival of their research activities.

For 'intellectual persuaders', engagement with the public is mainly expedient – a means to an end, which is usually the pursuit of intellectual work on the 'authoritative' or 'intellectual' pattern. The public needs persuading – of the validity and value of intellectual knowledge products – because public support is increasingly necessary to support intellectual work. This includes financial support: as taxpayers who finance government and research council transfers to research and teaching institutions, customers who buy products from corporations that sponsor research, and buyers of the books, articles and films intellectuals make. It also includes a willingness to tolerate intellectual work that has an increasing impact on society's security and welfare – from the potentially inflammatory tracts of a controversial historian to the potentially inflammatory torus of a particle physicist. Even if their conclusions are demonstrably correct, those being paid to think must convince the public (and its political representatives) that this is the best use of their time and of other people's resources. The imperative to persuade grows even stronger when – as their own methodology usually dictates – intellectual discoveries are only provisional, open to challenge, and are often seen to dispose of more certain knowledge than they create. Persuasion can be especially important for natural sciences and engineering, whose research can have significant impacts on the natural, social and political environment and often requires substantial public funding. In some instances (for example, the Lovelock and Sheldrake programmes noted earlier), the use of peer approval to project ideas to the public is supplemented by the use of public approval to pitch new ideas to intellectual peers.

A second group, which might be termed 'dialogical intellectuals', treat engagement with the public as a direct input into knowledge creation. For example, philosophers, influenced by the pragmatist rejection of a neutral algorithm for adjudicating ethical, aesthetic or scientific claims (Rorty 1999: xvi–xxxii), recognise that theirs is only one among a range of legitimate voices and they cannot discard laypeople's views. Some moral philosophers who once reasoned deductively over correct action in tragic-choice situations now work inductively, canvassing public opinion and trying to rationalise what is in the public mind (e.g. Hauser 2007). In a similar way, some economists who observe inconsistency between axiomatic decision theory and popular decision practice have ceased to ignore or re-educate the wider public, and instead rebuild their theories to accommodate its choices (e.g. Kahneman and Tversky 1979; Sonsino *et al.* 2002). Some cultural anthropologists, influenced by the reflexive turn in cultural anthropology (Marcus and Fischer [1986] 1999), make a concerted effort not to impose their own categories on to what is being studied but to develop dialogical knowledge so as to learn from others. Increasingly, scientific researchers and consultants involved in large public projects develop what some commentators have called 'dialogic democracy' (Giddens 1994: 115ff.): that is, they engage the public in environmental impact and risk assessments, recognising the validity of the layman's rationality over axiomatic rational choice theory (see also Latour 2004). Although not all of them have a distinct public profile in the traditional sense of the word, they engage publicly with the public they are studying so that the latter has a central input into the knowledge formation. Public engagement is promoted by information technology, enabling interaction that gives audiences a clearer role in developing and legitimating knowledge.

For dialogical intellectuals, engagement is more often an end in itself – motivated by belief that a public input into intellectual enquiry can usefully inform and direct it. Entering such dialogue, intellectuals deepen their knowledge through exposure to public opinion and action. Ideas extracted from this engagement may be made to challenge received intellectual ideas, tested for consistency on their own terms, or used to understand the often unintended social impact of intellectual thought. Increasing use of public opinion and survey research, oral history, media analysis, diary exercises and participant observation (among other methods that entail engagement) indicate a perceived need to receive more input from the public before ideas can be convincingly communicated about or to the public. Dialogue can be especially important for the social sciences and humanities, for which the public (or sections of it) are usually objects of research as well as targets for research output (and policies based on it). While some social analysts reacted to lay-audience challenges to the professional intellectual by seeking to renew an authoritative role, through axiomatic or abstract theorising, at least as many have reacted by embracing the dialogical role, launching new strands of empirical or participant-observer research in which what people think is a necessary component of what we think about people.

For both types of 'embedded' intellectual, the intellectual's original two missions, namely to create and disseminate knowledge, are recognised as increasingly hard to pursue without engagement in a third – the mobilisation of public input into research and teaching. The relentless expansion of universities, once viewed as entrenching what Halsey calls 'the donnish dominion' (Halsey 1982: 216), has made intellectual work a costly enterprise that the public must be called upon to finance, while also equipping the public to be more sceptical of intellectuals' declarations and more insistent that they prove their worth. High education levels and an increasing sense of the contested nature of knowledge claims make it difficult for intellectuals to rely on a vertical relationship of authority vis-à-vis their audience. Dissenting voices undermine the assertion of authority through professional expertise. Instead, intellectuals have to develop 'horizontal authority', enrolling a wider public in the production of knowledge – as financial or ideational contributors – at the same time as communicating new ideas to it.

The public intellectual and the intellectual public

Accepting public input to the methods and conclusions of intellectual thought, implying comparable if not identical status of 'lay' and 'expert' opinion, is a radical departure from the working methods of authoritative and professional intellectuals. Instead of claiming (through ascription or achievement) superior knowledge to the public, and approaching it in the fashion of legislator or teacher, intellectuals increasingly view knowledge as gaining substance or legitimacy only through interaction with the public mind, through a process of dialogue and persuasion. Instead of lecturing, embedded intellectuals have to argue their case to an audience that can question and criticise their arguments.

Two long-running developments, sometimes lumped together in the concept of 'knowledge society', drive intellectuals from top-down pronouncement into dialogue. One is the increasing educational attainment of the general public, as the length of compulsory schooling increases and a rising proportion moves from schooling into higher education, or equivalent professional development. The other is the increasing specialisation of intellectual work. This leads to a broadening of the public mind and a focusing of the intellectual mind, promoting meetings of the two.

An increasingly educated public is more resistant to being talked down to, and more inclined to demand a voice in conversations involving professional intellectuals. This is not to say that expansion of general and higher education has narrowed the knowledge gap between intellectuals and the general public. On the contrary, this gap has almost certainly widened. Academic journals (for instance, the *Economic Journal* or the *American Sociological Review*) are rarely as understandable to the 'educated lay reader' today as 50 or even 20 years ago. The epistemic distance between intellectual and lay conversation has been lengthened by increasingly technical use of language (especially mathematical and statistical), and increased use of referencing to past contributions, a knowledge of which is required in order to make sense of new contributions. What narrows as a result of expanding education is the evaluative distance between intellectuals and the public. 'Lay'-audience members become more competent at assessing the nature, coherence and effectiveness of intellectual arguments, and more confident in expressing scepticism or demanding clarification. This increases the public inclination to challenge, reserve judgement on or even outrightly reject intellectual arguments, without any claim to have received or fully understand the technical details of those arguments.

Education leaves 'lay' readers and listeners better equipped – or believing themselves to be better equipped – to assess the structure and coherence of intellectual arguments, without grasping their full content. The separation of evaluation from technical understanding is assisted by the enhanced formalisation and empirical testing of arguments that accompanied the rise of the professional intellectual. Formalisation involves the conversion of verbal arguments into models – making explicit the axioms or assumptions underlying those models, which a lay public can consider and challenge even without understanding the intricacies of the modelling that follows. (The 'professional' models can also often be reduced to a skeletal form which the public can come closer to understanding.) So, for example, many 'lay' people challenge complex economic models on the basis of assumptions that wages adjust to create full employment in labour markets, or that individuals and firms make rational maximising choices. Empirical testing confronts models with data, whose provenance and accuracy can be judged by a lay public independently of their assessment of the models. So, for example, climate change sceptics have challenged global warming models on the basis that their calibration uses data that can be alternatively interpreted, or whose accuracy can be disputed. Education can confer the ability (and confidence) to identify and challenge the style of intellectual argument, even when the argument's technical contents are beyond lay comprehension.

An increasing specialisation of intellectual work also increases the scope, and incentive, for an educated public to criticise the essence of a 'professional' argument without necessarily grasping all the details. Intellectuals' increasingly narrowly focused areas of enquiry can make them incapable of grasping the wider picture, so that the public – with a stronger understanding of its own interests – has (or claims) a role in judging the value of intellectual work. To the extent that intellectuals transmute into experts – 'knowing more and more about less and less' – they cede a role for independent evaluation by educated lay observers, who can (or believe they can) better view the context and significance of their work, knowing less about more. Once they begin to associate the growth in specialist knowledge with a decline in the ability to situate and evaluate this knowledge, members of the public (and their political representatives) become more confident in interrogating the specialists, and fit their knowledge into a more coherent, comprehensive natural or social worldview. Half a century ago, this view of intellectuals as disabled by their specialism led US congressional representatives to treat some of the most powerful scientific minds as comparable to those of children, unable to grasp the implications of their discoveries, or to accept adult regulation (Hall 1979). Across the Atlantic, Churchill (typically more pithily) pronounced that scientific experts should be "on tap but not on top". 'Declinists' have tended to argue that intellectuals yield to this tendency, perhaps even welcoming it – rarely resisting the loss of a public role, and often accepting confinement to narrow expertise, not least because the criteria for intellectual respectability have shifted from generalism to specialism.

Our argument is that any such confinement was only a short phase after the maturing of the professional intellectual, extensively reversed as intellectual activity becomes re-embedded in a more knowledgeable society. The reversal has been driven by new incentives for public engagement, frequently associated with new strands of thought that lend themselves to broad-based application. Natural and social scientists have increasingly succeeded in reversing the Churchillian assessment, transcending specialism to supply an overall vision, and taking a more direct role in the management, application and dissemination of their research by displacing former intermediaries. They have been incentivised to do so, as observed earlier, by the persuasive need to mobilise support for their work and the dialogical imperative to involve the public in it. This (re)assertion of a public dimension has been assisted by the power of specific discoveries to drive general conclusions, defying the suspicion that specialist reductionism would erode general understanding and undermine grand theory. The possibility of building a wider vision on deeper knowledge has been reinvigorated by the reconstruction of 'macro' mechanisms and structures on more detailed 'micro' foundations, in the form of physical and chemical explanations for the origins of the universe and of life, and evolutionary or rational-choice theories of their subsequent development.

Conclusion

We have argued that the thesis of decline in the incidence and influence of public intellectuals is – even in the Anglo-American context in which it has mainly been developed – deficient in at least three ways. First, it mistakenly presents the rise to dominance of the academy, as a centre (across the world) for the generation and dissemination of knowledge, as antithetical to public intellectual life. Second, it interprets external challenges to that dominance, and the increasing substitution of 'dialogical' for 'vertical' engagement with the public, as a weakening of the public intellectual role, when the opposite is the case. Third, it fails to recognise that the erosion of foundationalism increases the incentive for intellectuals and knowledge producers to engage with the public, as does the increased need for public support (legitimative and financial) of intellectual activity.

We have counterposed the declinist thesis by identifying three different forms of intellectual, with contrasting incentives for intellectual engagement. Times of change in the relative frequency of these forms have often been identified with decline, when they actually lead to widespread renewal of public engagement, though with different motivations and manifestations. We have further argued that a recent social re-embedding of intellectual activity, linked to an increasingly knowledgeable society, has led to the emergence of new forms of public intellectual engagement: for persuasion (reflecting a 'democratisation' in the allocation of resources to intellectual work), and for dialogue (reflecting a similarly democratic participation in the way that some knowledge, especially of the social world, is constituted and validated).

The emergence of 'embedded' intellectuals does not mean the eclipse of 'authoritative' and 'professional' intellectuals, nor an erosion of the scope and incentives for these longer established forms also to become publicly engaged. While it is true that intense professionalisation makes it more difficult for authoritative intellectuals to emerge and survive, it is perfectly compatible with the rise of the professional public intellectual. The professionally focused 'specific' intellectual may, far from becoming disengaged from the wider social relevance and impact of their work, actually become more strongly connected to them and motivated to pursue them, through ever closer linkage between their professional knowledge application and their practical living and working situation (Foucault 1980).

Two motivations for public engagement, initially characteristic of authoritative intellectuals, may be seen to persist. Intellectuals may use public fora (including broadcasting, the popular press, public enquiries and the internet) to disseminate and popularise intellectual understanding, and exercise 'thought leadership'. Philosophers Cyril Joad and Bertrand Russell were doing so on British radio in the 1950s, and the tradition continues today with (for example) David Attenborough presenting natural history documentaries and Nouriel Roubini posting his economic forecasts on the internet. Some intellectuals also enrol the public in their quest for knowledge, but purely as subjects for observation or agents of observation. For example, astronomers, including Carl Sagan, successfully engaged people's home computers in the Search for Extra-Terrestrial Intelligence, and naturalists collect data from mass observations such as the UK's 'Big Garden Birdwatch'.

But further forms of public engagement are also observable, of the 'horizontal' variety that is distinct from intellectual work after the high-point of professionalisation. Intellectuals now engage persuasively with the public in order to maintain support for their activity, because of its dependence on public funding and public goodwill. Intellectuals have grown in number and influence owing to public funding of research and teaching, and the employment security of academic tenure, afford them opportunities not available when they were reliant on personal fortunes, philanthropy or aristocratic tutoring. To persuade the public that such endeavours are worth supporting they must convince people that knowledge has value, in itself or through its practical applications – and that such applications as stem cell research, nuclear engineering and financial market deregulation genuinely promise more good than harm. While some intellectuals seek to persuade the public of the merits of their current projects, others mobilise opinion *against* the current orthodoxy, using popular belief in the relevance of a concept to counterweigh peer dismissal. In the most radical form of 'horizontal' engagement intellectuals involve the public in the actual production of knowledge. The public is sometimes involved directly by isolating popular belief and action as a form of knowledge, as in public opinion polling or the construction of decision models using commonly observed methods of risk evaluation. More often, the public is involved indirectly through mechanisms that allow it to shape the conditions for knowledge production, or pass judgement on its outputs – via, for example, referendums or 'citizens' juries' on proposals for technology adoption, and opinion surveys which indicate whether predicted trade-offs between present pain and future gain (on public spending or carbon emissions cuts) are politically deliverable.

The coexistence of vertical and dialogical engagement shows that the arrival of the persuasive intellectual has not meant the disappearance of the authoritative intellectual. Nor, despite the intra- and extra-institutional challenges noted earlier, has the professional intellectual gone into retreat. There is still thriving demand (and supply) for, among many other cases, statisticians as expert witnesses at a trial or public hearing, toxicologists as advisers at a public enquiry into waste dumping, and economists on a central bank's monetary committee. While these engagements with non-academic fora require the subjection of intellectual argument to non-peer review, they contribute to legitimising the intellectual process and demonstrating its social value (Beecher-Monas 2007). In addition, there is still demonstrable scope for intellectual reputation gained in a specialist area to propound specific views (or world-views) beyond it, as when Noam Chomsky attacks American foreign policy or Richard Dawkins condemns organised religion. Our principal contention has been that the rise of the professional intellectual has been wrongly interpreted as the eclipse of the classically defined intellectual by the 'expert', and that the subsequent arrival of the intellectual persuader has been wrongly interpreted as the dethronement or abdication of the public intellectual. Conversely, the arrival alongside traditional authority of intellectual professionals and persuaders has created new opportunities and motivations for public intellectual engagement. This has, in particular, enabled an increasing number of natural scientists to escape the pigeonhole of 'expert' for the wider intellectual airspace. The ranks of inwardly focused experts may have outgrown those who engage a wider public, but there has been no fall in the numbers who are publicly engaged. The number of publicly prominent scientists may have outgrown that of engaged intellectuals from the arts and humanities; but far from signalling decline, this results from new entry due to significant expansion of the public intellectual space.

Note

1 We thank Joel Isaac for his useful comments.

Bibliography

Agger, Ben (2001) *Public Sociology: From Social Facts to Literary Acts.* Lanham, MD: Rowman & Littlefield.

Argyris, Chris (1976) 'Single-loop and double-loop models in research on decision making.' *Administrative Science Quarterly* 21(3): 363–375.

Bauman, Zygmunt (1991) *Legislators and Interpreters: On Modernity, Postmodernity, and Intellectuals.* Cambridge: Polity Press.

Beck, Ulrich (2009) *World at Risk.* Cambridge: Polity Press.

Beecher-Monas, Erica (2007) *Evaluating Scientific Evidence.* Cambridge: Cambridge University Press.

Boggs, Carl (1993) *Intellectuals and the Crisis of Modernity.* Albany: State University of New York Press.

Bourdieu, Pierre (1988) *Homo Academicus.* Cambridge: Polity Press.

—— (1991) *Political Ontology of Martin Heidegger.* Cambridge: Polity Press.

—— (1996) *State Nobility; Elite Schools in the Field of Power.* Cambridge: Polity Press.

—— (2000) *Pascalian Meditatians.* Cambridge: Polity Press.

Brooks, David (2001) *Bobos in Paradise: The New Upper Class and How They Got There.* New York: Simon & Schuster.

Burawoy, Michael (2005) 'For public sociology.' *American Sociological Review* 70: 4–28.

Callon, Lascoumes and Barthe (2009) *Acting in an Uncertain World; An Essay on Technical Democracy.* Boston, MA: MIT Press.

Camic, Charles (1983) *Experience and Enlightenment: Socialization for Cultural Change in Eighteenth-Century Scotland.* Chicago, IL: University of Chicago Press.

—— (1987) 'The making of a method: A historical reinterpretation of the Early Parsons.' *American Sociological Review* 52(4): 421–439.

—— (1957) *Syntactic Structures.* The Hague: Mouton.

—— (1965) *Aspects of the Theory of Syntax.* Cambridge, MA: MIT Press.

—— (1983) *The Fateful Triangle: The United States, Israel and the Palestinians.* Boston, MA: South End.

—— (2003) *Hegemony or Survival.* New York: Metropolitan Books.

Collins, Randall (1998) *The Sociology of Philosophies: A Global Theory of Intellectual Change.* Cambridge, MA: Harvard University Press.

Corbyn, Zoe (2010) 'UEA broke FoI law.' *Times Higher Education,* 28 January. Online at www.timeshighereducation.co.uk/story.asp?storycode=410209 (accessed 25 May 2010).

Dawkins, Richard ([1976]1989) *The Selfish Gene.* Oxford: Oxford University Press (2nd edn).

Delanty, Gerard (2001) *Challenging Knowledge: The University in the Knowledge Society.* Buckingham: Open University Press.

Djilas, Milovan (1957) *The New Class: An Analysis of the Communist System.* New York: Harcourt Brace Jovanovich.

Dyer, Clare (2005) 'Professor Roy Meadow Struck off.' *British Medical Journal* 331: 177 (23 July). Available online at www.bmj.com/cgi/content/full/331/7510/177 (accessed 26 May 2010).

Etzioni, Amitai (2001) *The Monochrome Society.* Princeton NJ: Princeton University Press.

Etzioni, Amitai and Bowditch, Alyssa (eds) (2006) *Public Intellectuals; An Endangered Species.* Maryland: Rowman & Littlefield.

Evans, Mary (2005) *Killing Thinking: The Death of Universities.* London: Continuum.
Eyerman, Ron (1994) *Between Culture and Politics: Intellectuals in Modern Society.* Cambridge: Polity Press.
Florida, Richard (2002) *The Rise of the Creative Class.* New York: Basic Books.
Foucault, Michel (1980) 'Truth and power.' In: *Power/Knowledge*, ed. C. Gordon. New York: Pantheon.
Friedman, Milton and Rose, D. (1980) *Free to Choose: A Personal Statement*, New York: Harcourt Brace Jovanovich.
Friese, Heidrun and Wagner, Peter (1998) 'More beginnings than end. The other space of the university.' *Social Epistemology* 12 (1): 27–31.
Furedi, Frank (2004) *Where Have All the Intellectuals Gone? Confronting 21st Century Philistinism.* London: Continuum.
Galbraith, John Kenneth (1967) *The New Industrial State.* London: Penguin.
—— (1977) *The Age of Uncertainty.* London: BBC.
Gibbons, Michael, Limoges, Camille, Nowotny, Helga, Schwartzman, Simon, Scott, Peter and Trow, Martin (1994) *The New Production of Knowledge.* London: Sage.
Giddens, Anthony (1994) *Beyond Left and Right; The Future of Radical Politics.* Cambridge: Polity Press.
Gross, Mattias (2007) 'The unknown in process; dynamic connections of ignorance, non-knowledge and related concepts.' *Current Sociology* 55 (5): 742–759.
Hall, Harry S. (1979) *Congressional Attitudes towards Science and Scientists.* New York: Arno Press.
Halsey, A.H. (1982) 'The decline of donnish dominion?' *Oxford Review of Education* 8(3): 215–229.
Halsey, A.H. (1992) *The Decline of Donnish Dominion.* Oxford: Clarendon Press.
Hamilton, William D. (1964) 'The genetical evolution of social behaviour.' *Journal of Theoretical Biology* 7: 1–16 and 17–52.
Harrabin, Roger (2010) Getting the message, BBC News Channel, 29 May. Available online at http://news.bbc.co.uk/1/hi/science_and_environment/10178454.stm (accessed 30 May 2010).
Hauser, Marc (2007) *Moral Minds.* New York: Little, Brown.
Hawking, Stephen (1988) *A Brief History of Time.* New York: Bantam Books.
Hitchens, C. (2008) How to be a public intellectual. *Prospect* 146. Available online at www.prospectmagazine.co.uk/2008/05/howtobeapublicintellectual/ (accessed 12 May 2010).
Jacoby, Russell (1987) *The Last Intellectuals; American Culture in the Age of Academe.* New York: Basic Books.
Kadish, Alon (1993) 'Marshall and the Cambridge Economics Tripos.' In: *The Market for Political Economy*, eds A. Kadish and K. Tribe. London: Routledge, pp. 137–161.
Kahneman, Daniel and Tversky, Amos (1979) 'Prospect theory: an analysis of decision under risk.' *Econometrica* 47: 263–291.
Kerr, Clark (1963) *The Uses of the University.* Cambridge, MA: Harvard University Press.
Konrad, George and Szelenyi, Ivan (1979) *Intellectuals on the Road to Class Power.* New York: Harcourt Brace.
Latour, Bruno (2004). *Politics of Nature: How to Bring the Sciences into Democracy.* Cambridge, MA: Harvard University Press.
Lovelock, James (2000) *Homage to Gaia: The Life of an Independent Scientist.* Oxford: Oxford University Press.

Marcus, George and Michael Fischer ([1986] 1999) *Anthropology as Cultural Critique.* Chicago, IL: Chicago University Press (2nd edn).

Marshall, Alfred (1920) *Principles of Economics.* London: Macmillan (8th edn).

Marshall, George (2009) 'Leaked email climate smear was a PR disaster for UEA.' *Guardian,* 23 November. Available online at www.guardian.co.uk/environment/cif-green/2009/nov/23/leaked-email-climate-change (accessed 25 May 2010).

Nisbet, Robert ([1970] 1997) *The Degradation of the Academic Dogma.* Princeton, NJ: Transaction Publishers (2nd edn).

Popper, Karl ([1963] 2002) *Conjectures and Refutations.* London: Routledge.

Posner, Richard (2001) *Public Intellectuals: A Study in Decline.* Cambridge, MA: Harvard University Press.

Rorty, Richard (1979) *Philosophy and the Mirror of Nature.* Princeton, NJ: Princeton University Press.

Rorty, Richard (1991) *Philosophical Papers, Volume 1: Objectivity, Relativism and Truth.* Cambridge: Cambridge University Press.

Rorty, Richard (1998) *Philosophical Papers, Volume 3: Truth and Progress.* Cambridge: Cambridge University Press.

Rorty, Richard (1999) *Philosophy and Social Hope.* Harmondsworth: Penguin.

Rouanet, Sergio (2003) 'Religion and knowledge.' *Diogenes* 50(1): 37–50.

Samuelson, Paul A. (1948) *Economics.* New York: McGraw-Hill.

Sheldrake, Rupert (1987) *A New Science of Life.* London: Paladin.

Shellard, Dominic (2003) *Kenneth Tynan: A Life.* New Haven, CT: Yale University Press.

Smithson, Michael (1987) *Ignorance and Uncertainty: Emerging Paradigms.* New York: Springer.

Snow, C.P. (1959) *The Two Cultures: and A Second Look.* Cambridge: Cambridge University Press.

Sonsino, Doron, Benzion, Uri and Mador, Galit (2002) 'The complexity effect on choice with uncertainty – experimental evidence.' *Economic Journal* 112: 936–965.

Thompson, E.P. (1963) *The Making of the English Working Class.* London: Gollancz.

—— (1985) *The Heavy Dancers.* New York: Pantheon.

Young, James O. (2001) *Art and Knowledge.* London: Routledge.

Index

Not all authors cited are listed in the index. Readers requiring complete lists of cited authors and texts should consult the reference lists at the end of each chapter.

Printed in Great Britain
by Amazon.co.uk, Ltd.,
Marston Gate.